HUMAN FACTORS AND WEB DEVELOPMENT

Second Edition

HUMAN FACTORS AND
WEB DEVELOPMENT
Second Edition

Edited by

Julie Ratner

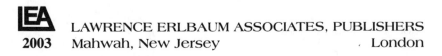

LAWRENCE ERLBAUM ASSOCIATES, PUBLISHERS
2003   Mahwah, New Jersey   London

Senior Acquisitions Editor:	Anne Duffy
Editorial Assistant:	Kristin Duch
Cover Design:	Kathryn Houghtaling Lacey
Textbook Production Manager:	Paul Smolenski
Full-Service Compositor:	TechBooks
Text and Cover Printer:	Sheridan Books, Inc.

This book was typeset in 10/12 pt. Times, Italic, Bold, Bold Italic.
The heads were typeset in Americana, Italic and Bold.

Lawrence Erlbaum Associates, Inc., Publishers
10 Industrial Avenue
Mahwah, New Jersey 07430

Library of Congress Cataloging-in-Publication Data

Human factors and Web development / edited by Julie Ratner.—2nd ed.
      p.   cm.
    ISBN 0-8058-4221-7—ISBN 0-8058-5225-5
    1. Web sites—Design.   2. Web sites—Psychological aspects.   3. World Wide
Web—Psychological aspects.   4. Human-computer interaction.   I. Ratner, Julie.

TK5105.888 .H86   2002
025.04—dc21   2002010673

Books published by Lawrence Erlbaum Associates are printed on
acid-free paper, and their bindings are chosen for strength and
durability.

Printed in the United States of America
10 9 8 7 6 5 4 3 2 1

Contents

Preface

When Chris Forsythe, Eric Grose and I co-edited the first edition of Human Factors and Web Development, our intent was to compile the definitive "seminal" research on the impact of the World Wide Web (WWW) development in the fields of cognitive psychology, engineering, user interface design, and educational technology.

In 1996, at the time we were planning the book chapters, there were very few publications or web specific resources for practitioners or researchers to cite, so we sought out the trend setters in academia, corporate, and government institutions. The purpose of the first edition was to share research on human factors, including user interface (UI) design standards, as best practices shifted to accommodate the delivery of information on the web.

As sole editor of the second edition, my approach to planning the content of this volume changed significantly because of the new technological landscape in 2001 and the global integration of the Internet in schools, libraries, homes, and businesses. Because many computer users are connected both at home and at work, the web has transformed communication, consumption patterns, and access to businesses, politicians, and neighbors halfway around the world. With numerous books on the user-friendly web sites, my challenge was to deliver a second edition with another radically different snapshot of the research being conducted at the beginning of the twenty-first century.

The chapters included in this book provide many answers to critical questions and propose thought-provoking research ideas for the future. Corporate and academic practitioners and their doctoral fellows and graduate students from around the globe all contributed their time, expertise, and unique research practices to collectively provide another comprehensive resource that our growing community of human factors and web development experts can share. My hope is that the chapters in this book will inspire leading edge research on accessible web applications or other topics in the coming years.

Julie Ratner, Ph.D.
Molecular, Inc.

Contributors

Johan Aberg, Linkopings Universitet

John R. Anderson, Carnegie Mellon University

Brian P. Bailey, University of Minnesota

Randolph Bias, Austin Usability, Inc.

Mon-Chu Chen, Carnegie Mellon University

Yun-Maw Cheng, University of Glasgow

Ed H. Chi, Xerox PARC

Mary P. Czerwinski, Microsoft Research

Laura J. Gurak, University of Minnesota

Marc Hassenzahl, User Interface Design GmbH

Chris Johnson, University of Glasgow

Joseph A. Konstan, University of Minnesota

Luke Kowalski, Oracle Corporation

Masaaki Kurosu, National Institute of Multimedia Education

Kevin Larson, Microsoft Research

Jonathan Lazar, Towson University

Deborah J. Mayhew, Deborah J. Mayhew & Associates

Robert Opaluch, Verizon Labs

Julie Ratner, Molecular, Inc.

Andrew Sears, UMBC

Nahid Shahmehri, Linkopings Universitet

Myeong-Ho Sohn, Carnegie Mellon University

Pawan R. Vora, Xor, Inc.

I

Digital Strategy: Planning in the 21st Century

1

Introduction

Deborah J. Mayhew
Deborah J. Mayhew & Associates

THE WEB AS SOFTWARE

The World Wide Web (WWW) began in 1989, the brainchild of Tim Berners-Lee at the European Organization for Nuclear Research (CERN), to address the need for a collaborative tool to help scientists share knowledge. It was then offered to the entire Internet community for experimentation, and thus the WWW was born (Horton, Taylor, Ignacio, & Hoft, 1996, p. 13).

What is the WWW? It is a huge, ever growing collection of hyperlinked documents (and now applications) created by independent authors and stored on computers known as Web servers. These documents are made accessible to anyone over the Internet via software applications called browsers and via various search engines accessible through browsers. Browsers provide a relatively user-friendly interface (at least compared to trying to use the Internet in the days before browsers), for navigating and viewing the total information and application space represented by all sites on the WWW. Thus, whereas the Internet is an electronic communications network that generally supports various kinds of communications, including e-mail, the WWW represents a repository of public information and transactions created by the public and accessible to the public via the Internet.

By analogy, the Internet is an electronic pony express, providing an electronic means to move information from one physical location to another. E-mail is an online post office/postal service, providing an online process for using the pony express (the Internet) to send and receive private mail. The WWW is an online library, storing and providing public access to a huge body of information (and now transactions as well), and search engines and browsers are partly an online Dewey decimal classification system, providing an online method for finding and viewing information and services stored on the WWW over the Internet.

Traditional public libraries, however, are very different from the WWW. On the WWW, anyone can publish a book (a Website) and add it to the library (the WWW). There are no publishing companies screening and editing books for publication and no librarians deciding what and what not to add to a library's collection. Because screening, editing, publishing, and purchasing intermediaries do not exist in the world of the WWW, prospective authors themselves must judge what content is useful and will be used and what presentation is effective. WWW users must wade through the overwhelming morass of unscreened, unedited published information to find useful, credible, well-presented information and services.

One of the WWW's great powers is its openness. It represents the ultimate in freedom of speech and makes huge, geographically dispersed audiences easily available to any aspiring author, publisher, vendor or service provider at very low cost. One of its great weaknesses, on the other hand, is its lack of quality control, both for content and for presentation. Again, this role has traditionally been assumed by publishing companies and libraries.

If it somehow became possible and easy for anybody to self-publish a book and distribute it to any library, two new services would probably quickly emerge— one to help author/publishers produce high-quality publications and another to help library users find, screen, and evaluate publications. In the case of the WWW, which makes self-publication easy, search engines and browsers help users find publications and services, and a well-established and completely applicable service is, in fact, readily available to help prospective author/publishers produce high-quality publications and services—the field of software usability engineering.

Software usability engineering, or software human factors, is a maturing field, now over 20 years old. It has produced useful methods and guidelines for designing usable software—methods and guidelines that are directly applicable to the design of Websites.

There are many reasons why an individual or organization might decide to publish a Website. These include (but are certainly not limited to) attracting business or offering a service or product; sharing a special skill, knowledge, or perspective; giving specific answers whenever users need them; shrinking distances and reducing isolation of all kinds; supporting learning and intellectual stimulation and pursuing personal interests (Horton et al., 1996, pp. 3–6).

Because these reasons are very broad, the range of Website authors and publishers is also broad. Many are unaware of the field of usability engineering and what it can offer them in their efforts to produce a useful, usable Website. One purpose of this book is to introduce potential or current Web authors and publishers to this field to show its relevance to Web design.

Reasons for browsing the WWW are just as broad and varied, including the obvious, usual reasons, such as finding information about products and services, shopping, researching a topic, accessing up-to-date information about topics such as stock markets, and finding people or organizations with similar interests. Some less usual reasons include finding the birth parents of an adopted child, getting information on wanted criminals, getting online access to comics, and finding information about colleges and universities.

Many, if not most, WWW users are discretionary; many have had relatively little computer experience and may even be computer phobic. Many are only occasional users, and many may not speak English as their native language (the Web is indeed worldwide). Thus, usability issues in Web design seem even more critical than they are for traditional software application design.

Consider the following interesting statistics that have undoubtedly changed since this writing but can serve as examples:

- There are over 476 million Internet users worldwide.
  (http://taskz.com/DesignData/general_user_indepth.htm)
- At least 55% of Internet users are non-English speakers.
  (http://taskz.com/DesignData/general_user_indepth.htm)
- In the United States, at least 52% of at-home Internet users are women.
  (http://taskz.com/DesignData/general_user_indepth.htm)
- 52% to 76% of online customers abandon their shopping carts before completing online purchases.
  (http://taskz.com/DesignData/general_user_indepth.htm)
- In mainland China, 7.2 million households have access to the Web.
  (http://www.treese.org/intindex/00–06.htm)
- 90% of full-time, 4-year college students in the United States use the Internet.
  (http://www.treese.org/intindex/00–02.htm)
- There are over 3,119,000 unique public Websites. (http://wcp.oclc.org)
- Forty-seven percent of unique public Websites are U.S.-based—no other single country accounts for more than 5%. (http://wcp.oclc.org)
- Of the top 50 sites most frequently linked from other sites, amazon.com is #9, cnn.com is #11, weather.com is #17, mapquest.com is #22 and whitehouse.gov is #34. (http://wcp.oclc.org)
- Two thirds of U.S. small businesses are Internet users.
  (http://www.nua.ie/surveys)
- At least fifteen percent of U.S. senior citizens are Internet users.
  (http://wcp.oclc.org)

- The number of African American Internet users increased 19% between 2000 and 2001. (http://wcp.oclc.org)
- Half of U.S. Hispanics are now Internet users. (http://wcp.oclc.org)

Clearly the Web is here, and here to stay.

Initially, Websites were read-only information sites, and the only interaction supported on Websites was navigation to other parts of a site or to other sites. In addition, tools for developing Websites were limited to providing ways to define content; presentation of content was mostly outside the control of individual Website authors and under the control partly of browser applications and partly of users through the browser. Tools now exist to support the development of interactive Websites that are almost indistinguishable from traditional types of interactive software and to provide extensive support for design and presentation to Web authors.

Even when designing simple, informational, read-only Websites, designers can draw on usability principles from the field of traditional software usability engineering. Much traditional software user interface (UI) design involves effective presentation of information and the design of effective navigational aids and methods. These issues are common to both traditional software application and traditional Website design. As Websites become more interactive and new tools and technologies provide more design options to developers, the principles, guidelines, and methods of software usability engineering become even more directly applicable. A random tour around the WWW, however, shows that the knowledge and wisdom collected in the field of software usability engineering over the last 20 years is generally not being applied to Website design.

Some useful distinctions between Web design and traditional software design include the following (Nielsen, 1996):

- On the Web, response times are generally slower and less predictable than on traditional software, and this places significant limits on practical file sizes and interaction in general.
- The WWW has fewer restrictions on style, and no interaction guidelines have yet emerged comparable to the interaction style guides available to traditional software developers.
- Web users, unlike most users of traditional software, are mainly discretionary users, a fact that increases the importance that Web software be "walk-up-and-use."
- The WWW is huge, much larger than any application, with fluid boundaries between sites. It has an increased need for navigational support and sense of place in the user interface.

Yet these differences are not qualitative, but more a matter of degree. They hearken back to the days of mainframe computers and dumb terminals, when designers were similarly constrained by poor performance, the absence of style guides, and major restrictions on display capabilities. Today, many applications are intended for an audience of discretionary, generally computer-naive users. Similarly, automation today is moving away from discrete, nonintegrated applications toward suites of separate, but tightly integrated, applications that share databases and common UIs. As Web development tools become more capable, traditional UI issues arise, such as

- The importance of good conceptual model design to communicate structure, support user models, and facilitate effective navigation (Nielsen, 1996).
- The importance of context information to maintain a sense of place, both global (i.e., on the WWW) and local, (i.e., in a site; Nielsen, 1996).
- The importance of good screen design.
- The importance of effective feedback.
- The importance of user-centered design and usability testing (Web design is interaction design, as are other types of software design) (Nielsen, 1996).

Thus, the WWW does not call for a new field to support designers. The Web is software, really just a different platform, and the field of software usability engineering can be directly applied to Web design and development. This book reinforces this view by offering examples of the application of usability engineering methods and principles to Web design.

APPLYING THE USABILITY ENGINEERING LIFECYCLE TO WEB DEVELOPMENT

The Usability Engineering Lifecycle (Mayhew, 1999) documents a structured and systematic approach to addressing usability within the product development process. It consists of a set of usability engineering tasks applied in a particular order at specified points in an overall product development lifecycle.

Several types of tasks are included in the Usability Engineering Lifecycle, as follows:

- Structured usability requirements analysis tasks.
- An explicit usability goal-setting task, driven directly from requirements analysis data.
- Tasks supporting a structured, topdown approach to user interface design that is driven directly from usability goals and other requirements data.
- Objective usability evaluation tasks for iterating design toward usability goals.

Figure 1.1 represents in summary, visual form, the Usability Engineering Lifecycle. The overall lifecycle is cast in three phases: requirements analysis, design/testing/development, and installation. Specific usability engineering tasks

8

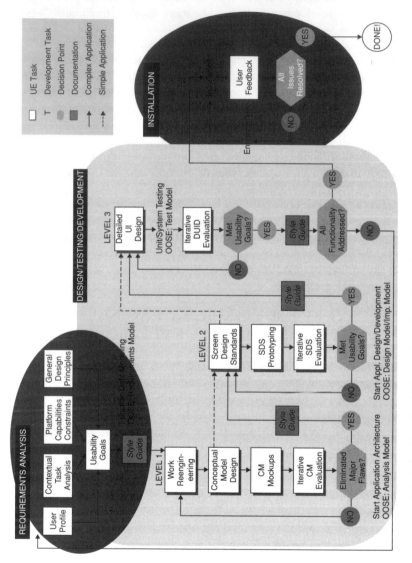

FIG. 1.1. The Usability Engineering Lifecycle. *Note.* Taken from Mayhew (1999).

within each phase are presented in boxes, and arrows show the basic order in which tasks should be carried out. Much of the sequencing of tasks is iterative, and the specific places where iterations would most typically occur are illustrated by arrows returning to earlier points in the lifecycle.

In considering how to adapt the Usability Engineering Lifecycle to a Web development project, the distinction between *tasks* and *techniques* is an important one.

A usability engineering *task* can be defined as an activity that produces a concrete work product that is a prerequisite for subsequent usability engineering tasks. Each task has some conceptual goal that defines it. For example, the goal of the user profile task is to gain a clear understanding of those characteristics of the intended user population that will have a direct bearing on which design alternatives will be most usable to them.

A *technique*, on the other hand, is a particular process or method for carrying out a task, and for achieving a task goal. Usually there are a number of alternative techniques available for any given task. For example, for the task of user profile, alternative techniques include distributing user questionnaires and conducting user or user manager interviews. Generally, techniques vary in how costly and time-consuming it is to execute them, in the quality and accuracy of the work products they generate, in how difficult they are for nonspecialists to learn and use, in the sophistication of the technology required to carry them out, and in other ways.

The key to the general applicability and flexibility of the Usability Engineering Lifecycle lies in the choice of which techniques to apply to each task, not in the choice of which tasks to carry out. All the tasks identified in the lifecycle should be carried out for every development project involving an interactive product in order to achieve optimal usability. However, the approach to any given project can be adapted by a careful selection of techniques based on project constraints.

Each usability task in the overall Usability Engineering Lifecycle is briefly described below, with special emphasis on adapting the tasks to a Web development project.

Phase One: Requirements Analysis

User Profile

A description of the specific user characteristics relevant to user interface design (e.g., computer literacy, expected frequency of use, level of job experience) is obtained for the intended user population. This will drive tailored user interface design decisions and also identify major user categories for study in the contextual task analysis task, later described.

The problems of doing a user profile for a Website or application are similar to those of doing one for a vendor company: The users are not easily accessible—or even known. However, a user profile is still doable.

You can get help from marketing to identify and access potential users. You can do a "quick and dirty" user profile based on interviewing marketing personnel or others who may have contact with potential users. And, you can at least solicit user profile information after the Website or application is implemented, through the site itself, and use this information to update and improve the new versions of the Website and to build new related Websites and applications. You could create a link that takes users to a user profile questionnaire and provide some incentive (discounts, raffle entries, etc.) for them to fill it out and send it back.

Contextual Task Analysis

A study of users' current tasks, work-flow patterns, and conceptual frameworks is made, resulting in a description of current tasks and work flow and an understanding and specification of underlying user goals. These will be used to set usability goals and drive work reengineering and user interface design.

The problems of doing a contextual task analysis for a Website or application are, again, very like those of doing one for a vendor company: The users are not easily accessible or even known, and the "work" may not currently be performed by intended users. However, a contextual task analysis is still doable.

In a contextual task analysis for a Website or application, the focus might be more on what people want or need, rather than on how they currently do tasks. You can often get help from marketing to identify and access potential users. You *can* do a contextual task analysis of average people doing personal tasks at home, such as catalog ordering, planning travel, buying a new car, and other activities. After the fact, you can also solicit task-related information from the Website itself: You could have a feedback page and use feedback to update and maintain a Website, and to build new related Websites and applications.

Usability Goal Setting

Specific, qualitative goals reflecting usability requirements extracted from the user profile and contextual task analysis, and quantitative goals defining minimal acceptable user performance and satisfaction criteria based on a subset of high-priority qualitative goals, are developed. These usability goals focus later design efforts and form the basis for later iterative usability evaluation.

In most cases, at least when designing public Websites or applications, ease of learning and remembering goals will be more important than ease-of-use goals, due to the infrequency of use. Most users do not visit a given Website daily, and many often visit a site only once.

Ease of navigation and maintaining context will usually be very important qualitative goals for Websites and applications.

When formulating quantitative performance goals, Web designers need to be aware that system response time will limit and impact user performance, and that system response time will vary enormously, depending on the users' platforms.

In many cases of Website or application design, relative quantitative goals may be appropriate (e.g., "It must take no longer to make travel reservations on the Website than it does with a travel agent by phone").

Platform Capabilities and Constraints

The user interface capabilities and constraints (e.g., windowing, direct manipulation, and color) inherent in the technology platform chosen for the product (e.g., Apple Macintosh, MS Windows, or product-unique platforms) are determined and documented. These will define the scope of possibilities for user interface design.

Unlike some of the other Usability Engineering Lifecycle tasks, the platform capabilities and constraints task will often be more, rather than less, complicated when designing a Website or application, as compared to designing traditional software applications. This is because normally, with the exception of the case of some intranet applications, designers have to assume a potentially very large number and wide variety of hardware and software platforms. Internet users' platforms will vary, possibly widely, in at least the following ways:

- Screen size and resolution.
- Modem speed.
- Browser capabilities (varies by vendor and version), for example:
  - Controls available through the browser (vs. must be provided within the site or application).
  - Browser interpreters (e.g., version of HTML, Java).
  - Installed "helper applications" or "plug-ins" (e.g., multimedia players).

Web user interface designers need to design for the expected range of platform capabilities and constraints. For example, one common technique is to have a control at the entry point to a Website or application that allows users to choose between a "graphics mode" and a "text mode." Thus, users with slow modems can "turn off" any graphics that would seriously degrade download time and view an alternative text-only version of the site or application.

Similarly, if a Website or application requires specific helper applications or plug-ins, many are now designed to allow immediate downloading and installation of the required helper application or plug-in. The user interface to downloading and installing helper applications or plug-ins is often not very user friendly, but at least providing the capability is a step in the right direction.

In general, Web designers need to be aware that if they take full advantage of all the latest Web capabilities, many users will find their Website or application unusable. Care needs to be taken to provide alternative interfaces for users with lower end platforms.

General Design Guidelines

Relevant general user interface design guidelines available in the usability engineering literature are gathered and reviewed. They will be applied during the design process to come, along with all other project-specific information gathered in the above tasks. Most general software user interface design principles and guidelines will be directly applicable to Website and application design. Things to bear in mind that do make designing for the Web a little different from designing traditional software include

- Response times are slower and less predictable on the Web, limiting what design techniques are practical.
- There are no existing user interface standards for the Web.
- Browsers and users, rather than designers and developers, may control much of the appearance of Web content.
- Web users may be mainly discretionary and infrequent, increasing the need for "walk-up-and-use" interfaces.
- The World Wide Web is a huge and fluid space with fuzzy boundaries between sites. There is thus an increased need for navigational support and a "sense of place."

These differences are not quantitative, however. Rather, they are a matter of degree. Web platforms simply place more constraints on designers than do traditional platforms. Designing for the Web is somewhat like designing for traditional software 20 years ago, although the capabilities on the Web are now catching up fast.

Phase Two: Design, Testing, and Development

Level 1 Design

Work Reengineering.   Based on all requirements analysis data and the usability goals extracted from it, user tasks are redesigned at the level of organization and work flow to streamline work and exploit the capabilities of automation. No user interface design is involved in this task, just abstract organization of functionality and work-flow design. This task is sometimes referred to as information architecture.

Sometimes you are actually simply engineering—rather than reengineering—work, as your Website or application supports "work" unlike anything most of the intended users currently do (e.g., deciding on the structure for an information space users were not previously able to access). Nevertheless, you still could have done a contextual task analysis to discover users needs and desires, and could base your initial work organization on this analysis.

In most cases, even when a particular job is not currently being done, something highly related to that job is being done, and this can be the focus of a contextual task analysis. In addition, once an initial release of a Web application is in production, you can perform another contextual task analysis to discover how it is being used and where it breaks down, and use these insights to reengineer the underlying work models for later releases. Also, just as when designing traditional software, you can still validate your reengineered work models empirically with evaluation techniques.

Conceptual Model Design. Based on all the above tasks, initial high-level design rules are generated for presenting and interacting with the application structure and navigational pathways. Screen design detail is not addressed at this design level.

The conceptual model design is equally as important in Website and application design as in traditional software design. A conceptual model design for a Website might typically include rules that would cover the consistent presentation of:

- Location and presentation of site title/logo.
- Panes (e.g., for highest level links, context information, or page content).
- Links to different levels in the site map.
- "You Are Here" indicators on links.
- Links versus other actions (e.g., "Submit").
- Links versus nonlinks (e.g., illustrations).
- Inter- versus intrasite links.
- Inter- versus intrapage links.

On very simple Website or application projects, it may not be necessary to separately document the conceptual model design. Nevertheless, the conceptual model must be explicitly designed and validated.

Conceptual Model Mock-ups. Paper-and-pencil or prototype mock-ups of high-level design ideas generated in the previous task are prepared, representing ideas about how to present high-level functional organization and navigation. Detailed screen design and complete functional design are not in focus here.

Instead of paper foils or throw-away prototypes, in the case of Websites and applications the mock-ups can simply be partially coded products, for example, pages, panes, and navigational links with minimal page content detail.

Iterative Conceptual Model Evaluation. The mock-ups are evaluated and modified through iterative evaluation techniques such as formal usability testing, in which real, representative end users attempt to perform real, representative tasks with minimal training and intervention, imagining that the mock-ups are a real product user interface. This and the previous two tasks are conducted in iterative cycles until all major usability "bugs" are identified and engineered out of the Level 1 (i.e., conceptual model) design. Once a conceptual model is relatively stable, system architecture design can commence.

Remote usability testing is particularly well suited to testing Websites and applications. Software products are available that allow a tester and a user to share control of a screen while talking on the phone. This can replace traditional usability testing in which the tester and user are side by side in the same location, and is more practical when users are widely dispersed geographically.

Level 2 Design

Screen Design Standards. A set of product-specific standards and conventions for all aspects of detailed screen design is developed, based on any industry and/or corporate standards that have been mandated (e.g., Microsoft Windows, Apple Macintosh, etc.), the data generated in the requirements analysis phase, and the product-unique conceptual model design arrived at during Level 1 design. Screen design standards will ensure coherence and consistency—the foundations of usability—across the user interface.

Screen design standards are just as important and useful in Web design as in traditional software design. Besides the usual advantages of standards, in a Website they will help users maintain a sense of place within a site, as your site standards will probably be different from those on other sites.

On very simple Website or application projects, it may not be necessary to formally document the screen design standards. Nevertheless, the screen design standards must be explicitly designed and validated.

Web design techniques (both good and bad) tend to be copied—perhaps other Web designers will copy your screen design standards! Perhaps someday, we will even have a set of universal Web screen design standards supported by Web development tools, not unlike Microsoft Windows and Apple Macintosh standards. This would contribute greatly to the usability of the Web, just as the latter standards have done for traditional software.

Screen Design Standards Prototyping. The screen design standards (as well as the conceptual model design) are applied to design the detailed user interface to selected subsets of product functionality. This design is implemented as a running prototype. Instead of paper foils or throwaway prototypes, in the case of Websites and applications, the prototypes can simply be partially coded products, for example, selected pages, panes, and navigational links, now with complete page content detail.

Iterative Screen Design Standards Evaluation. An evaluation technique such as formal usability testing is carried out on the screen design standards prototype, and then redesign/reevaluate iterations are performed to refine and validate a robust set of screen design standards. Iterations are continued until all major usability bugs are eliminated and usability goals seem within reach. Again, remote usability testing can be particularly useful when testing Websites and applications.

Style Guide Development. At the end of the design/evaluate iterations in Design Levels 1 and 2, you have a validated and stabilized conceptual model design and a validated and stabilized set of standards and conventions for all aspects of detailed screen design. These are captured in the document called the product style guide, which already documents the results of requirements analysis tasks. During detailed user interface design, following the conceptual model design and screen design standards in the product style guide will ensure quality, coherence, and consistency—the foundations of usability.

For simple Websites and applications, as long as good design processes and principles have been followed, documentation may not be necessary. For complex Websites—even intranet Websites—with many designers, developers and/or maintainers of a constantly evolving site, documenting requirements analysis work products and design standards is very important, just as it is on large, traditional software projects.

Level 3 Design

Detailed User Interface Design. Detailed design of the complete product user interface is carried out based on the refined and validated conceptual model and screen design standards documented in the product style guide. This design then drives product development.

For simple Websites or applications, designers might bypass documenting user interface design at the conceptual model design and screen design standards levels and simply prepare detailed user interface design specifications directly from standards they have informally established at these earlier design levels. Developers can then code directly from these specifications.

For more complex Websites or applications, the conceptual model design and screen design standards should usually be documented before this point. Then, developers can code directly from a product style guide or detailed user interface design specifications based on a product style guide prepared by the user interface designer.

Iterative Detailed User Interface Design Evaluation. A technique such as formal usability testing is continued during product development to expand evaluation to previously unassessed subsets of functionality and categories of users, and also to continue to refine the user interface and validate it against usability goals.

On projects developing relatively simple Websites and applications, it might be more practical to combine the three levels of the design process into a single level, where conceptual model design, screen design standards, and detailed user interface design are all sketched out, in sequence, before any evaluation proceeds. Then a single process of design and evaluation iterations can be carried out.

In this case, an iterative detailed user interface design evaluation will be the first usability evaluation task conducted. Thus, evaluation must address all levels of design simultaneously. This is practical only if the whole application is fairly simple, which Websites and applications often are. In addition, it is important to remember that even if detailed user interface design is drafted before any evaluation commences, it is still crucial to consider all the same design issues that arise in the conceptual model design and screen design standards tasks when conducing design in a three-level process.

In Websites and applications of intermediate complexity, the first two design levels (conceptual model design and screen design standards) might be combined into a single process of design and evaluation iterations to validate them simultaneously, and then an additional process of design and evaluation can be carried out during the detailed user interface design level. Alternatively, Level 1 can be carried out with iterative evaluation, and then Levels 2 and 3 can be collapsed into one iterative cycle. In either case, this will be the second usability evaluation cycle conducted and can indeed focus mainly on screen design standards and detailed user interface design, as the focus will have been conceptual model design during an earlier evaluation task.

Also, in the case of Website or application design, mock-ups, prototypes, and application code can all simply be final code at different points of completion rather than paper foils, throwaway prototypes, and then final code. As in previous design levels, remote usability testing can be particularly useful when testing Websites and applications.

Phase Three: Installation

User Feedback.   After the product has been installed and in production for some time, feedback is gathered to feed into enhancement design, design of new releases, and/or design of new, but related, products.

User feedback can be solicited directly from a Website or application. This can be done by providing a link on the site taking users to a structured feedback page, or by offering direct e-mail from the site and asking users to provide free-form feedback. You can even have survey questions pop up, triggered by specific usage events. An advantage of this kind of solicited feedback is that it collects feedback while the experience is fresh in the user's mind.

You might need to provide some incentive for users to take the time to provide feedback, especially if you provide a structured form and it is at all lengthy. Possible incentives include entry in a raffle or discounts on products or services.

The user feedback techniques that lend themselves most easily to Websites and applications include questionnaires and usage studies. Other techniques (interviews, focus groups, and usability testing) are more difficult to employ because they require the identification and recruitment of users to meet in person with project team members, something that may not be difficult on intranet sites but may be next to impossible on Internet sites.

To a young Web designer or developer who launched a career in the Internet age and has worked primarily on Web development projects, the Usability Engineering

Lifecycle may at first blush seem much too complex and time-consuming to be practical in the fast-paced world of Web development. And, if you consider only traditional and rigorous techniques for lifecycle tasks, and typical time frames for the development of very simple read-only Websites, this is a fair assessment. For example, whereas I have often conducted task analyses techniques that took several months to complete and formal usability tests that took a month or more, I have also worked on Web development projects that from start to launch took a total of 8 to 12 weeks. Clearly, one cannot spend several months conducting task analyses—just one of the first steps in the Usability Engineering Lifecycle—when the whole project must be completed in 2 to 3 months!

Two points must be made, however. First, the Usability Engineering Lifecycle is highly flexible and adaptable through the selection of techniques applied to each task and the collapsing of design levels, and can accommodate even projects with very limited time frames. I have, in fact, successfully adapted it to even 8-week Web development projects.

Second, initially Websites were functionally very simple compared to most traditional software applications, and so the fact that they typically took 8 to 12 weeks to develop, as compared to months, or even years, for traditional software applications, made some sense. Now, however, Websites and applications have gotten more and more complex, and, in many cases, are much like traditional applications that happen to be implemented on a browser platform. The industry needs to adapt its notion of reasonable, feasible, and effective time frames (and budgets) for developing complex Web-based applications that simply are not the same as simple content-only Websites. This includes adapting its notion of what kind of usability engineering techniques in which to invest.

In a December 2000 report by Forrester Research, Inc., called "Scenario Design" (their term for usability engineering), it is pointed out that

Executives Must Buy Into Realistic Development Time Lines and Budgets
The mad Internet rush of the late 1990's produced the slipshod experiences that we see today. As firms move forward, they must shed their misplaced fascination with first-mover advantage in favor of lasting strategies that lean on quality of experience.
   • **Even single-channel initiatives will take eight to 12 months.** The time required to conduct field research, interpret the gathered information, and formulate implementation specs for a new Web-based application will take four to six months. To prototype, build, and launch the effort will take another four to six months. This period will lengthen as the number of scenarios involved rises.
   • **These projects will cost at least $1.5 million in outside help.** Firms will turn to eCommerce integrators and user experience specialists for the hard-to-find experts, technical expertise, and collaborative methodologies required to conduct Scenario Design. Hiring these outside resources can be costly, with run rates from $150K to $200K per month. This expenditure is in addition to the cost of internal resources, such as project owners responsible for the effort's overall success and IT resources handling integrations with legacy systems."

I agree that 8 to 12 *months* is a more realistic time frame to develop a usable Website or application that will provide a decent return on investment (ROI). And, if this is the overall project time frame, there is enough time to use traditional usability engineering techniques to more reliably ensure Website usability, a major contributor to ROI. Depending on the complexity of a Website or application, a budget somewhere between $100,000 and $250,000 should pay for a rigorous and thorough usability engineering program. This is a small fraction of the $1.5 million estimated by Forrester for all the outside help a Website sponsor will need.

SUMMARY

The well-established field of software usability engineering, or software human factors, can be directly applied to the design of Websites, but, unfortunately, as in the early days of traditional software development, this knowledge base is not yet being routinely applied during Website development. Emerging lingo among Web users indicates the general lack of usability of current Websites. The following examples were published in *Wired* and have been anonymously passed around via e-mail:

> *Dorito syndrome:* Feelings of emptiness and dissatisfaction triggered by addictive substances that lack nutritional content. "I just spent 6 hours surfing the WWW and now I've got a bad case of Dorito syndrome."
> *Crapplet:* A badly written or profoundly useless Java applet. "I just wasted 30 minutes downloading this stinkin' crapplet!"
> *World wide wait*: The real meaning of WWW.
> *Link rot*: The process by which links on a Web page become obsolete as the sites they are connected to change location or die.
> *CobWebsite*: A Website that has not been updated for a long time; a dead Web page.

In an article about the Internet, Kraut (1996, p. 35) wrote that usability once on the WWW is one issue, but another issue is the usability of the process of getting connected to the Internet in the first place: "Connecting to the Internet is still an overwhelming hurdle for many families. As many online service providers are discovering, walking naive users through the vagaries of modems and networking software is expensive, but necessary."

I remember only too well my own experience. I resisted getting on the Internet for years because, in spite of cheery reassurances from technically inclined clients, colleagues, and friends, I knew that it was not going to be easy, and I am not one to put up with unusable technology. Of course I was right. In spite of being computer literate, it took me a whole month to get connected. First, I had to purchase various pieces of critical software. Then, when more software was required and supplied by my access provider, I could not successfully download it, and my provider could not explain why. When I finally had the software installed, the provider sent me a long manual specifying a set of meaningless numbers that had to be entered correctly in 20 or 30 dialog box fields with meaningless labels. I followed the instructions scrupulously, and things still did not work. Again my provider seemed just as mystified as I was. Finally, after many phone calls, wasted hours and frustration, not to mention outrage at the lack of usability of the connection process, I finally got onboard. Now, of course, I cannot live without nor do business without the Internet. But I felt a disturbing sense of déjà vu throughout the whole process of getting connected, as if I had been transported back about 20 years ago in terms of the usability of the technology. Clearly, the potential of the Internet and the WWW will never be realized until the well-established principles and methods of software usability engineering are systematically applied to this new technology.

The remaining chapters of this book offer insights into general and specific principles and guidelines applicable to Web design, as well as details of specialized techniques that can be applied in the overall context of the Usability Engineering Lifecycle to achieve usability in Web design.

We hope that this book finds its way into the hands of the right people. We hope that usability engineers find this book helpful to understand how their expertise can be applied to Web design and that they are inspired to get involved in any Web development in their client or employer organizations. We hope that Web designers of all kinds find the book a good introduction to the usability issues involved in Web design and an inspiration to pursue other usability engineering resources to support their efforts. We hope that it does not take another 20 years for the Web development community to catch up to the traditional development community in terms of the application of usability engineering principles and methods to product development, and we hope that his book contributes to the process.

(Portions of this chapter were excerpted or adapted from the book *The Usability Engineering Lifecycle* (Mayhew, 1999), used with permission. Other portions were excerpted or adapted from an article by Deborah H. Mayhew that first appeared on http://taskz.com/ucd_discount_usability_vs_gurus_indepth.htm in September 2001.)

REFERENCES

Foster, R. W. (1996). The internet: A perspective for internetworking. *Inside DPMA, 1*, 3.

Horton, W., Taylor, L., Ignacio, A., & Hoft, N. L. (1996). *Web page design cookbook*. New York: Wiley.

Kraut, R. (1996). The Internet@Home. *ACM Communications, 39*, 12, 32–35.

Mayhew, D. J. (1999). *The Usability Engineering Lifecycle*: San Francisco: Morgan Kaufmann Publishers.

Mayhew, P., Discount Usability vs. Usability gurus: a Middle grund, www.taskz.com/ucd_discount_vs_gurus_indepth.php

Nielsen, J. (1996). Jakob Nielsen on Web site usability. *Eye for Design, 3(4)*, 1, 6–7. (User Interface Engineering)

Sondereqqet, P. (2000). Scenario Design, Forrester Report, Cambridge, MA.

2

Universal Usability
and the WWW

Andrew Sears
UMBC

INTRODUCTION

In the context of information technologies, universal usability has been defined as "having more than 90% of all households as successful users of information and communications services at least once a week" (Shneiderman, 2000). Many issues relevant to this broad definition are beyond the scope of this chapter. As a result, we shift the focus from having a specific percentage of all households successfully using the World-Wide Web (WWW) on a regular basis to ensuring that the information and services provided via the WWW are usable for every citizen regardless of the environment in which they are located or the technologies they are using. Given this focus, our goal is to provide practitioners and researchers with insights into the issues that must be addressed, some solutions that exist, and pointers to key resources when designing for

- Individuals with perceptual, cognitive, and physical impairments and disabilities.
- Specific user groups (e.g., children and the elderly).
- An international audience (e.g., cultural and language issues).
- Varied technologies (e.g., network connection and screen size).

Unexpected benefits often appear as the needs of specific populations are addressed. A classic example is the curb cut (the ramp between a road and sidewalk). The original goal was to allow wheelchair users to cross streets more easily, but many other individuals also benefit, including individuals pushing baby carriages and riding bicycles. Not only do curb cuts provide benefits for many individuals beyond the intended users, the amount of material required is reduced by including curb cuts in the original plans. Other examples include the telephone, television subtitles, and speech transcription systems (all originally developed to assist deaf people). Various audio recording technologies (e.g., multitrack tape recorders and cassette tape recorders) were developed to assist people who are blind. Many more examples are discussed by Newell (1995). Of particular interest is how guidelines developed to address the needs of individuals with various impairments or disabilities can be applied to other situations. In fact, many of these guidelines prove useful when designing for children, the elderly, and international audiences as well as for devices with small displays (e.g., PDAs and cell phones). These guidelines may also prove useful when individuals experience situationally-induced impairments or disabilities (SIIDs). Examples include hearing impairments created by noisy environments, physical impairments created by cold or moving environments, and visual impairments created by poor lighting. Of course, SIIDs are different in that they tend to be temporary and dynamic, suggesting that individuals experiencing SIIDs are likely to develop different strategies for accommodating the difficulties they experience. These issues are also important given the trend toward increased legislation requiring information technologies to be accessible to individuals with impairments and disabilities (e.g., GSA, 2002).

This chapter focuses on issues relevant to the delivery of information and services via the WWW. More specifically, we focus on issues that can be addressed by the individuals who make these resources available. As a result, user-based solutions (e.g., alternative input devices) are not addressed. In the sections that follow, we discuss the issues involved. We also provide some basic guidelines and pointers to additional resources.

Perceptual, Cognitive, and Physical Impairments and Disabilities

The World Health Organization (WHO) defines an impairment as a loss or abnormality of body structure or function (WHO, 2000). Examples include a loss of muscle power (e.g., a weak grip), reduced mobility of a joint (e.g., difficulty bending fingers), and uncontrolled muscle activity (e.g., shaking hands). Impairments can be caused by health conditions such as arthritis, spinal cord injuries, or Parkinson's disease. Impairments can also be caused by the context in which the interaction takes place (e.g., environment or tasks). For example, working in a cold environment can result in a loss of muscle power and reduced mobility, much like arthritis.

The WHO defines a disability as a difficulty an individual may have in executing a task or action (WHO, 2000). Therefore, a disability is associated directly with some activity, such as interacting with the WWW. Impairments (e.g., reduced range of mobility in the fingers) may, or may not, result in a disability (e.g., difficulty using the keyboard and mouse). Disabilities can also be caused by the context in which the interaction takes place. An example would be working in a noisy environment that makes it difficult to hear audio output from a computer speaker.

Therefore, perceptual, cognitive, and physical impairments and disabilities can be associated with an underlying health condition or with the situation in which the interactions are occurring (Sears & Young, 2003). Health-induced impairments and disabilities (HIIDs) differ from situationally-induced impairments and disabilities in that they tend to be more stable and permanent. As a result, individuals experiencing HIIDs may develop different strategies than individuals experiencing SIIDs (e.g., Keates, Clarkson, & Robinson, 2000; Sears, Karat, Oseitutu, Karimullah, & Feng, 2001).

Perceptual Impairments and Disabilities

Perceptual impairments and disabilities include difficulties associated with the four fundamental senses: vision, hearing, smell, and touch. In the context of the WWW, difficulties with vision are often the primary concern. However, given the increased use of audio, issues associated with hearing difficulties must also be considered. Smell and touch are not addressed in this chapter. For a more comprehensive discussion of perceptual impairments and disabilities, refer to Jacko, Vitense, and Scott (2003).

Visual Impairments and Disabilities

Visual abilities fall along a continuum. Given the availability of corrective devices (e.g., glasses and contact lenses), visual abilities are typically assessed after correction using a variety of measures, including visual acuity, contrast sensitivity, field of view, and color perception. Visual acuity measures the ability to distinguish fine detail (Kline & Schieber, 1985), contrast sensitivity assesses the ability to detect patterned stimuli at various contrast levels (Wood & Troutbeck, 1994), the field of view measures the area over which effective vision is maintained (Kline & Schieber, 1985), and color perception assesses the ability to discern and identify color within the useful field of view (Kraut & McCabe, 1994). Three broad groups are discussed below: individuals who are blind, individuals with low vision, and individuals with impaired color perception.

Blind computer users cannot make effective use of visual displays. Individuals who are blind typically access information on computers using either text-to-speech (TTS) technologies or braille displays. At present, both technologies can be used to provide access to textual information, but graphical information is typically not accessible. Ongoing research may provide blind users with improved access to graphical aspects of computer interfaces in the future.

Given existing technologies, the following guidelines should be considered to help ensure the accessibility of WWW-based materials for blind computer users:

• At present, effectively accommodating blind users requires that all information be made available in text. Textual descriptions must be provided for all graphics, applets, videos, and any other nontextual item that would normally require the user to view the screen. These textual descriptions should not only provide a name for the object, but should also allow the user to understand the information that is conveyed through the nontextual item. For example, the name of an image may serve as a link to a longer textual description of that image. These issues become even more critical when the nontextual information is used for navigation. Although image maps allow users to click on part of a graphic to navigate to new information, they introduce new barriers for blind users.

• Auditory descriptions may prove useful if the information is difficult to read when processed by standard text-to-speech technologies (e.g., mathematical formulas).

• Cursor positioning tasks can also prove problematic. Any interactions that normally require mouse-based input should also be supported via the keyboard. Keyboard-based navigation should be consistent, using the same keys throughout the interactions. Standard tab-based navigation should be tested to ensure that users move through the interface in a logical and expected order.

• Consider the demands placed on the user's working memory. Most information will be delivered using audio, but auditory information is transient. After listening to a list of ten options, users may no longer remember the early options. General guidelines should prove useful for designing auditory menus, such as those used when designing interactive voice-response systems. In addition, providing short descriptions that allow users to make informed decisions and careful information organization can prove useful.

Although it is important to consider blind users, the number of individuals with low vision far exceeds the number of individuals who are blind. Low vision refers to a wide range of impairments, including reduced visual acuity, contrast sensitivity, or field of view. Low vision can also refer to color perception difficulties, but these are addressed in a separate category.

In the context of the WWW, low vision becomes a concern when the individual can no longer interact effectively with traditional computer interfaces without alteration. Examples include individuals with extreme near- or far-sightedness, blurred vision, tunnel vision, or night blindness. Numerous computer-based tasks can be affected, including reading text, identifying feedback or small changes on the screen, and making use of alternative visual media (e.g., images and video).

Because individuals with low vision typically prefer to make use of their residual visual capabilities, solutions designed for individuals who are blind are unlikely to be widely accepted (Edwards, 1995a). At the same time, low vision encompasses a wide range of impairments, so the specific difficulties any individual may experience are likely to be unique. Basic guidelines include

• The most fundamental recommendation is to use a simplified visual design. All users will benefit if visual clutter is reduced, relevant information is highlighted, and all information that is displayed is carefully organized.

• Reading text can be particularly problematic. Ideally, users would be able to adjust all aspects of the text, including both the font and the size of the letters.

• Users may see only a fraction of the screen at any given time. This may be due to the use of screen magnification software or a reduced field of view. This has implications for information organization and navigation. It can also place additional demands on the users' working memory because they may need to scroll both vertically and horizontally before seeing all of the contents of the display. Finally, the inability to view the entire screen at one time can make visual effects more difficult to notice. Auditory feedback can prove useful, especially for alerting the user of important events.

• Many individuals with low vision benefit from increased contrast between the foreground and background. Consider the implications of high-contrast settings on backgrounds, graphics, and, perhaps more important, on the text that is displayed.

Individuals with difficulty perceiving colors are often referred to as color blind. Roughly 8% of men and 0.4% of women have difficulty perceiving colors. Most of these individuals who are color blind experience difficulty distinguishing two colors (e.g., red–green color blindness is most common; blue–yellow color blindness is less common). Relatively few cannot distinguish any colors and therefore only see shades of gray. Addressing the difficulties these individuals experience is further complicated by inappropriately adjusted monitors and highly variable lighting conditions.

Colors must be selected in such a way that individuals can still extract the necessary information even when the colors are not perceived as originally intended. A simple approach is to begin with a black-and-white (or gray-scale) design and then add color where necessary, ensuring that color is not the sole means of communicating information. When using multiple colors, select combinations that are easy to distinguish. Ideally, colors would vary in two or more dimensions (i.e., hue, saturation, or brightness), as this makes distinguishing colors easier. All of this must be done while remembering that users may customize their display (e.g., manually specify a foreground or background color), turn off image loading (e.g., when background images or icons are used), or access the material using a reduced color palette (e.g., 216 or fewer colors).

Hearing Impairments and Disabilities

Hearing impairments and disabilities are becoming increasingly important as the use of multimedia, video, and audio grows. Since it is becoming more common to access the WWW from nontraditional environments, the ability to use material in a noisy environment is becoming increasingly important. Allowing users to adjust the volume, pitch, and other characteristics of the acoustic information they receive can improve interactions. However, individuals with significant hearing impairments (e.g., individuals who are deaf or those working in noisy environments) may not be able to make effective use of information delivered acoustically regardless of these settings. For these individuals, redundant delivery mechanisms are critical. In general, it should be possible to deliver acoustic information visually as well. For example, visual effects can be substituted for attention-attracting sounds, closed captioning can be provided for video, and transcripts can be provided for any recorded speech.

Cognitive Impairments and Disabilities

Cognitive impairments and disabilities may be caused by disease, injury, or environmental or task characteristics such as stress, heat, or fatigue. Well-designed information technologies have the potential to enhance the quality of life and independence for individuals with cognitive impairments and disabilities. Effective information technologies can reduce social isolation and provide enhanced communication capabilities, intellectual stimulation, and greater independence while simultaneously allowing for appropriate monitoring and supervision (Newell, Carmichael, Gregor, & Alm, 2003).

In this section, we focus on the fundamental issues that must be addressed when designing for individuals with cognitive dysfunction in the context of the WWW. Cognitive capabilities fall along a continuum, the details of which are beyond the scope of this chapter. For a comprehensive discussion of these issues, refer to Newell, Carmichael, Gregor and Alm (2003). As with all development efforts, user involvement is recommended throughout the design process. User involvement will provide useful guidance, especially when designing for individuals with a specific set of impairments or disabilities. Unfortunately, when designing for the WWW this user population is often poorly defined. As a result, a different approach is required. Due to these conditions, we must understand the most common difficulties users may experience.

Language skills are highly variable and can be significantly affected by a variety of cognitive impairments and disabilities. Interfaces that utilize simplified vocabularies, avoid complex syntax, and avoid the use of metaphors will reduce the difficulties individuals with language difficulties may experience.

Spatial abilities may also be affected, indicating that simplified interface designs can prove beneficial. Carefully organizing information, guiding users to the information they need or the next step in a multistep process, and minimizing the amount of unnecessary text, graphics, and other elements that may distract the user can help.

Similarly, problem-solving skills may be affected by a variety of cognitive impairments and disabilities. The primary objective would be to minimize the use of problem-solving skills. If such skills are required, careful design based on the abilities of the intended users is required. The issues involved in addressing difficulties with memory and attention are explored in more detail below, followed by a brief collection of general design recommendations.

Memory

Cognitive impairments and disabilities often involve difficulties with both working memory and long-term memory. As with all interface design, it is best to minimize the demands that are placed on the users' working memory, but this becomes even more critical for individuals with cognitive dysfunction. Working memory becomes critical if the system presents information that must be used later in the interaction. Because the computer is presenting this information to the user, the first goal should be for the computer to supplement the users' working memory by storing this information and making it available when needed. Memory-support capabilities and effective error correction mechanisms can also prove useful. For example, difficulties with working memory can make navigation more difficult. Users may forget where they have been or what they are trying to locate. These users may benefit from effective history mechanisms, as well as tools that help them keep track of their goals.

Long-term memory becomes an issue when the necessary information resides with the user. In this case, the most important support may be to provide cues that may prompt the necessary memories. Computers can be used to supplement long-term memory by helping individuals store and organize information, but this is less relevant in the context of this chapter and is not discussed further. Some conditions, such as dementia, can result in difficulties remembering sequences of events and activities (i.e., episodic memory). Carefully designed navigational aids that help individuals remember where they have been and where they are going may prove particularly beneficial to individuals who experience difficulties with episodic memory.

Attention

Selective attention allows individuals to focus on relevant details while ignoring the irrelevant. Divided attention allows individuals to focus on multiple, simultaneous activities. Both selective and divided attention can be affected by cognitive impairments and disabilities. It is difficult to recommend a simple solution for situations where the users' attention must be divided among multiple activities, as each scenario may require different accommodations. It is best to avoid these situations whenever possible and to design carefully when they cannot be avoided. For example, additional visual and auditory feedback may be useful for attracting the users' attention at critical times. History mechanisms that allow users to review what they have done, where they are, and events that were missed may prove useful.

Difficulties with selective attention highlights one reason why complexity should be minimized. Increased complexity implies that more information is presented at one time, which implies that there is more information the user must ignore when focusing on any particular activity. Reducing complexity and careful use of attention-attracting visual effects can help individuals with selective-attention difficulties. This can be particularly relevant on the WWW, where numerous items frequently compete for the user's attention (e.g., banner ads and animated graphics).

Design Recommendations

Continuous involvement of potential users can help, but addressing the vast range of users that may be involved when designing for the WWW can be difficult. Given the complexity of cognitive impairments and disabilities, checklists cannot ensure effective solutions (Carmichael, 1999). At the same time, the following list of recommendations can provide a useful starting point:

• Complexity is perhaps the most significant problem when designing for individuals with cognitive impairments or disabilities. There are countless opportunities to reduce complexity. Options include simplifying the vocabulary and syntax of text, eliminating unnecessary graphic or textual information, simplifying graphics, limiting choices, and using a standardized layout and color scheme.

• Difficulties with selective attention can make it hard to focus on the relevant information. Options include simplifying the design, reducing the number of items that compete for attention, careful use of layout to highlight important information, effective use of attention-attracting visual and auditory cues to highlight relevant information, and even the use of multisensory feedback when information is particularly important. Difficulties with divided attention can make it difficult to attend to multiple simultaneous activities. The best advice is to avoid situations that require users to attend to two or more items simultaneously. If this cannot be avoided, consideration should be given to enhancing visual and auditory feedback when critical information becomes available or activities occur.

• Reduce the demands placed on the user's working memory. Options include providing tools to explicitly support the user's working memory, avoiding the use of long sequences of actions, and to not require users to unnecessarily remember information when moving from one part of the interface to another.

• Difficulties with navigation are related to memory but are worthy of a separate entry in this list. Certain memory difficulties can lead to significant problems with navigation, which is particularly important in the context of the WWW. Navigation should help users remember where they have been, where they are, and where they are trying to go. Using a standard layout throughout the entire site and providing high-level overviews that allow for immediate navigation to key locations on a Web site can also prove useful.

• Simplifying the vocabulary and syntax was already mentioned during the discussion of complexity. It is also a separate entry in this list due to the significance of this issue and the variety of implications it has on designing for the WWW. Simplifying the vocabulary and syntax is critical, and avoiding jargon and metaphors is also important.

• Graphics can help or hinder users. Simple graphics that convey useful information may be beneficial, whereas complex graphics may lead to confusion. Unnecessary graphics can distract the user, especially animated graphics. Redundant graphics accompanying text may convey information more effectively, but textual descriptions of these graphics should be provided for individuals who have difficulty understanding the images. Finally, the time required to download graphics, as well as the implications for individuals with visual impairments, should be considered.

• Transient information (e.g., audio) can result in additional difficulties because individuals with cognitive impairments and disabilities often process information more slowly than individuals without similar conditions. For example, transcripts may be useful supplements for video clips with audio. The rate, volume, pitch, and other properties of speech must be set appropriately, and mechanisms should be provided to allow this information to be reviewed at the user's own pace.

• Alternative interaction mechanisms must be considered, including the use of screen readers, screen magnification programs, and alternative input devices. This can affect the use of various options, such as columns, tables, and image maps. If screen readers may be used, many of the guidelines presented in the section on visual impairments and disabilities become relevant.

• Basic guidelines, including providing frequent feedback, helping users avoid errors and correct those that do occur, and guiding users through their tasks, become even more important.

Physical Impairments and Disabilities

In this chapter, we focus on a subset of physical impairments and disabilities that may hinder an individual's ability to physically interact with computing technologies (e.g., those affecting the upper body). In particular, we focus on the relationship between physical impairments and disabilities and an individual's ability to utilize the WWW. For a more comprehensive discussion, refer to Sears and Young (2003).

Physical impairments and disabilities can affect muscle power (e.g., paralysis and weakness), movement control (e.g., uncontrolled movements, tremors, and coordination difficulties), mobility (e.g., range and ease of motion of the joints), and physical structures (e.g., missing fingers). The underlying cause can be a health condition, an injury, the environment, or the activities in which an individual is engaged. Frequently, individuals with physical impairments or disabilities will make use of alternative input devices. Because the selection of an alternative device is under the user's control, it is beyond the scope of this chapter. However, it is important to note that these devices are typically designed to emulate the standard keyboard or mouse and frequently result in slower data entry.

Fundamental actions, as well as more complex actions, are often more difficult for individuals with physical impairments or disabilities. For example, Trewin and Pain highlight a variety of difficulties with basic activities, including pressing keys on a keyboard (e.g., accidentally pressing multiple keys, pressing the wrong key, or multiple presses of the same key), positioning the cursor, and clicking on objects (Trewin & Pain, 1999). More complex actions, including dragging objects and clicking the mouse button multiple times, were even more problematic. These difficulties are often compounded by increased sensitivity to both target size and movement distance (e.g., Casali, 1992; Radwin, Vanderheiden, & Lin, 1990).

While the connection to motor activities is clear, physical impairments and disabilities may also affect both cognitive and perceptual activities. These effects become apparent when analyzing both fundamental interactions and the high-level strategies adopted by individuals with physical impairments (PIs). For example, Keates, Clarkson, and Robinson (2000) provide data showing that individuals with PIs required 18% longer to complete basic cognitive activities and 25% longer for basic perceptual activities, whereas Sears et al. (2001) identified differences in the high-level strategies adopted by individuals with and without PIs when using the same speech recognition system to accomplish a variety of tasks. Differences in motor, cognitive, and perceptual activities must be considered.

Design Recommendations

As with cognitive impairments, a simple checklist cannot ensure effective solutions, but the following guidelines should be considered when designing for individuals with physical impairments or disabilities:

- Avoid device- or platform-specific capabilities. Users may use any number of input devices, output devices, and operating systems. Alternative input devices typically emulate the standard keyboard or mouse, suggesting that interactions should be based on the generic capabilities of standard devices.
- In general, keyboard use tends to be less problematic, as compared to mouse usage. This suggests that all interactions should be possible using only the keyboard. Whenever possible, this should include access to applets, image maps, and other controls.

- Keyboard-based navigation should be carefully designed to match user expectations.
- Basic actions (e.g., key presses and cursor positioning) will take longer. More complex actions (e.g., double-clicking and dragging objects) will be even more problematic.
- Cursor-control activities are more sensitive to both target size and distance the cursor must be moved. Larger targets can prove particularly useful, as can careful information organization that reduces the distance users must move the cursor.
- Additional time may also be required to perform cognitive and perceptual activities. As a result, alternative designs cannot be compared based only on the physical movements required.
- Avoid using multiple simultaneous key presses and repetitive key presses because these can be particularly problematic. Similarly, actions that require users to hold keys or buttons down while moving the cursor may be problematic and should be avoided.
- Consider the possibility that some interactions may be performed using a single hand, finger, or pointing stick.
- Speech-based interactions may prove useful, especially in the case of severe physical impairments (e.g., high-level spinal cord injuries).

Designing for Specific User Groups

Given the increasing number of individuals using the WWW, there are more individuals from well-defined user groups for which additional issues should be considered. We focus the issues involved in designing for two specific populations: children and the elderly.

Designing for Children

As Internet access continues to expand, both in private households and public facilities like schools and libraries, the Internet will become an increasingly important technology for children. In this section, we briefly summarize some of the more important issues that must be considered when designing for children. This is a challenging task given the rapid changes that can be observed in both motor and cognitive skills. For a more comprehensive discussion, refer to Bruckman and Bandlow (2003).

Very young children do interact with computers. For the youngest of children, this often involves little more than observing the screen and listening to sounds. As children grow older, interactions become more sophisticated. For example, children as young as 2 will interact with computers, but their motor skills, language skills, ability to focus on a single activity, and ability to solve problems is limited. With time, these skills mature, but the rate at which they change can vary dramatically. In general, by the time children reach approximately 7 years of age their language and motor skills have developed substantially, as has their ability to focus on a single activity. By the time children reach approximately 12 years of age, problem-solving skills have also matured, and interactions can be designed in much the same way as they would be for adults. The following should be considered when designing for children:

• As with all development efforts, the intended users should be involved throughout the development process. The same is true for systems designed for children. Contextual inquiry, technology immersion, and participatory design are all suggested. Some basic guidelines that can be applied regardless of the technique include keeping the experience informal, observing children in their own environment, avoiding time pressures, making it clear that you need their help by asking for their opinions and feelings, and making sure that any note taking is done as discretely as possible (Druin, 1998).

• Because fine-motor skills develop with age, children will be able to select larger targets more easily. Children also experience more difficulty with activities that require the mouse button to be held down, double-clicking, or multiple keys to be pressed simultaneously (e.g., Bederson et al., 1996; Inkpen, 2001; Joiner, Messer, Light, & Littleton, 1998). Note that distinguishing the left mouse button from the right may be problematic, suggesting that having left- and right-mouse buttons perform the same function may prove useful.

• The ability to attend to a specific activity also develops with age. Careful use of visual and auditory cues may help focus the child's attention. Animation can also prove useful. It is important to recognize that inappropriately timed feedback can also serve to distract the user from the intended activity. Rewards can be useful for younger children. An appropriate level of challenge may also help focus the child's attention.

• Reading skills must be considered when designing textual information. Limited data suggests that children prefer larger fonts (Bernard, Mills, Frank, & McKnown, 2001). For the youngest of children, text should be avoided entirely.

• Designs must consider the knowledge and experience of children. Difficulty distinguishing left from right is just one example. Others include limited experience with some metaphors that are frequently used in standard computer applications (e.g., folders of documents).

• Consider designs that allow complexity to increase gradually as the user becomes comfortable. Start by clarifying what is possible. Then, as children learn, additional information, functionality, and interaction opportunities can be provided. Altering the cursor as it moves over different objects can be effective for disclosing interaction opportunities.

• There are legal issues that must be considered when designing for children, including privacy, parental consent, and age-appropriate content. The Children's Online Privacy Protection Act (COPPA) of 1998 is an example of government legislation in this area (COPPA, 2001).

Designing for Older People

A growing segment of the population is aging at the same time as information technologies are becoming more pervasive. The WWW can provide a variety of benefits to older individuals, including improved access to information, enhanced communications capabilities, and increased independence. For more information about the aging process, refer to Rogers and Fisk (2000). For a more detailed explanation of the issues involved in designing information technologies for older users, refer to Czaja and Lee (2003).

Two general issues must be addressed when designing for older users. First, many of these users have limited experience using information technologies. This suggests that basic recommendations for designing for novice users may prove useful. Fortunately, while older individuals may have less experience and often encounter more difficulties using these technologies, they appear to be both willing and able to use computers. Czaja and Lee provide more information about the acceptance of technology by older individuals (Czaja & Lee, 2003). Second, the aging process is associated with changes in perceptual, cognitive, and motor functions. Whereas general trends do provide useful insights as we develop WWW-based materials for this growing segment of the population, we must also acknowledge that as with any group of individuals, the skills, knowledge, and experiences of older individuals will vary dramatically. In the following sections, we highlight several fundamental changes associated with aging that may affect interactions with information technologies.

Perceptual Abilities

Visual and auditory capabilities tend to deteriorate with age. Visual acuity, contrast sensitivity, color perception, and the ability to adapt to lighting changes may decline, so issues such as glare may be problematic with increased age. Basic activities, including reading text, identifying small targets, dealing with patterned backgrounds, perceiving differences in color, and dealing with displays that are visually complex, all become more difficult. Similarly, hearing abilities typically decline with age, resulting in difficulties with high frequencies, understanding speech, and locating the source of sounds (Schieber, Fozard, Gordon-Salant, & Weiffenbach, 1991). As a result, understanding recorded or synthesized speech may be more difficult, and sounds intended to attract attention may be missed. Recommendations for individuals with perceptual impairments and disabilities, especially those for individuals with low vision and hearing impairments, are important when designing for older users.

Cognitive Abilities

Age-related changes in cognition include difficulties with attention, working and long-term memory, and various problem-solving related processes (Park, 1992). Further, older individuals generally require longer to complete basic cognitive tasks. Although older users should not necessarily be considered disabled, the earlier recommendations for individuals with cognitive impairments and disabilities, especially those that address issues related to memory and attention, are important when designing for this population as well.

Motor Skills

As individuals age, motor skills often deteriorate. In particular, response times slow, involuntary movements become more common (e.g., tremors), voluntary movements become harder to control, and mobility impairments (e.g., difficulty bending fingers due to arthritis) become more common (Rogers & Fisk, 2000). As a result, older users tend to experience greater difficulty using standard input devices, including the keyboard and mouse. Examples include accidentally pressing keys multiple times and trouble with tasks that require fine positioning, double-clicking, or dragging actions using the mouse (e.g., Charness, Bosman, & Elliot, 1995; Riviere & Thakor, 1996; Smith, Sharit, & Czaja, 1999). Again, recommendations for individuals with physical impairments or disabilities can prove useful, especially those related to the importance of target size and avoiding more complex actions (e.g., multiple simultaneous key presses, double-clicking, and dragging).

Designing for Language and Cultural Differences

The Internet is inherently international. Resources made available via the Internet will be accessible by individuals who speak a variety of languages. In addition to speaking and reading different languages, these individuals will have varied cultural backgrounds. Therefore, the basic issues involved in developing software for international audiences also apply when developing materials for the WWW (e.g., Luong, Lok, Taylor, & Driscoll, 1995; O'Donnell, 1994; Sears, Jacko, & Dubach, 2000). At the same time, it is critical to acknowledge that truly effective solutions require more than just translating text from one language to another (Russo & Boor, 1993). The following issues must also be considered:

• *Text:* When text is translated from English to other languages, alternative fonts may be required, and the number of characters may increase substantially (Belge, 1995). This can affect the overall layout of the interface as well as the amount of horizontal and vertical scrolling required. Further, the translation process must address any use of jargon or domain-specific terminology. The order in which words are read may also differ. Although English is read from left to right and top to bottom, this is not true of all languages. This can affect the layout of text as well as the overall organization of the display (O'Donnell, 1994).

• *Date, time, currency, and number formats:* Most countries use the Arabic number system, but the formats used to display dates, times, and currencies differ, as does the use of decimals and commas.

- *Graphics, symbols, and icons:* The same graphic, symbol, or icon may convey fundamentally different concepts to individuals with different cultural backgrounds. Social norms determine image acceptability, and images that are acceptable in one culture may prove confusing or even offensive in another. In particular, any images depicting religious symbols, the human body, women, or hand gestures must be carefully evaluated (Russo & Boor, 1993).
- *Color:* Colors convey different meanings in different cultures and must be adjusted during the internationalization process. For example, red represents danger in the United States and happiness in China (Salomon, 1990).
- *Metaphors:* Metaphors typically build on a user's preexisting knowledge. Since backgrounds, experiences, and expectations differ, a metaphor that works well in one culture may be ineffective in another.

A general awareness of the issues involved is important when individuals from various countries may be accessing a site. Under these conditions, English is perhaps the most appropriate language for business-oriented sites. Many international users will have at least a rudimentary understanding of English, but the vocabulary should be limited to common words, and complex syntax should be avoided. Because English may not be the user's first language, the overall layout should be made more explicit to help guide users to the information they require. Some users may normally use a different format, but using well-defined formats for dates, times, currency, and numbers will reduce confusion. For the reasons highlighted above, it may be best to minimize the use of metaphors, colors, graphics, symbols, and icons that may be misinterpreted. When used, these elements should be evaluated carefully for each target audience. File sizes should be minimized, especially for graphics, because international users may experience longer download times, but ensuring that the proper message is conveyed is even more important.

When additional resources are available and a WWW site is intended for a specific set of markets, it may prove useful to create several versions of the site. Under these conditions, it is even more important to begin with a clear understanding of the intended users (e.g., backgrounds, preferences, and goals). Understanding these issues will provide guidance with respect to the changes that are necessary when customizing the site for different markets. Ideally, teams of experts in each target country would be identified early to allow for input throughout the development process. For a more detailed explanation of the issues involved in international and intercultural design, refer to Marcus (2003).

Designing for Varied Technologies

The WWW is now being accessed using desktop PCs, laptop computers, handheld devices, PDAs, and mobile phones. The processing speeds, available memory, screen sizes, network connections, and input mechanisms vary dramatically. Processing speeds continue to increase and memory continues to get smaller and less expensive, but utilizing these resources may lead to difficulties for users working with older or less powerful technologies. For example, any material that requires the full power of a desktop PC is less likely to function properly on a PDA or mobile phone without significant alteration. Similarly, designing material to utilize the latest capabilities of the newest version of a browser may make these materials inaccessible when individuals use older browsers.

While users can enter information at a reasonable pace using a standard keyboard and mouse, these devices are rarely available when using a PDA, mobile phone, or other handheld device. More often, limited input mechanisms exist, resulting in significantly slower data entry. At present few guidelines exist, but fundamental differences in processing speeds, available memory, and input mechanisms should be considered if users may be accessing the material being developed using anything other than a traditional desktop or laptop computer. Other differences, which have been explored in more detail, include the effects of network connection speed and screen size.

Network Connection Speed

Various studies have confirmed that the time it takes to download and display information can have a significant effect on user actions and perceptions (e.g., Jacko, Sears, & Borella, 2000; Ramsay, Barbesi, & Preece, 1998). Longer download times have been associated with documents being rated as less interesting (Ramsay et al., 1998) and confusion (Jacko et al., 2000). The delays users experience interact with the media they encounter, affecting their perceptions of the quality of the information being received, how well the information is organized within the site, the company's understanding of the issues involved in delivering information via the WWW, and how likely they are to recommend the site to others (Jacko et al., 2000). In general, longer delays will result in more negative feelings about the material received (see Jacko et al., 2000, for an exception regarding text-only sites). Research also suggests that some of the graphic enhancements designers employ to make a site appear more professional provide little benefit (e.g., animated graphics, graphic bullets, and graphic links, Jacko et al., 2000). However, these enhancements may prove useful if they directly contribute to users' ability to locate the information they need (Jacko et al., 2000). The designer's challenge is to balance the benefits users receive with the delays they experience.

Many factors contribute to these delays, but few can be controlled by the individuals responsible for developing and delivering information via the WWW (see Sears & Jacko, 2000, for a more detailed discussion). An organization can provide faster connection between the server and the Internet or a more powerful server, both of which can reduce delays. Similarly, end users can get faster network connections. Propagation and queuing delays are inherent to the Internet and largely beyond the control of any individual user or company. The only way the individuals designing the material can directly affect the delays users experience is by reducing the amount of information that must be transmitted. They can reduce the delays users perceive, as opposed to the actual delays they experience, by designing this information so it is displayed more quickly once it is received.

Although the general trend toward faster Internet connections continues, slower connections are still common. Many users have fast Internet connections at work, but the vast majority still use relatively slow connections when at home. Mobile devices tend to provide slower connections, the availability of certain technologies may differ depending on where the individual lives and works, and simple economics may make some technologies inaccessible for many individuals. As a result, designers must acknowledge the importance of network delays and understand the variety of network connections their users are likely to employ.

The designer's options are limited, but understanding these options can result in an improved experience for the user. After minimizing the amount of information that must be transmitted and ensuring that this information is displayed as quickly as possible, the designer must still evaluate the resulting materials to ensure that the delays users will experience are acceptable (e.g., Sears & Jacko, 2000; Sears, 2003). These issues are discussed in more detail below. While the recommendations below can be implemented today, several additional features may also prove useful if implemented (Shneiderman, 2000). Examples include allowing users to specify byte counts to limit the size of documents that are downloaded (a feature already implemented in some e-mail programs) or allowing users to specify whether they would like to download high-, medium-, or low-resolution versions of the graphics used in a particular document. For a more detailed explanation of the issues involved in designing network-based interactions, refer to Dix (2003).

Reducing the Effects of Network Delays

Two approaches can be used to reduce the effects of network delays. The first is to actually reduce the delays by minimizing the size of the documents users must download. The second is to design these documents such that they can be displayed more quickly. The following guidelines are important when designing for slower network connections:

• Some basics: Style sheets can prove useful because they reduce the amount of formatting in HTML documents. Use of include files, applets, and plug-ins should be minimized because these can result in additional delays. If users may download large files (e.g., audio/video, Word, and PDF), provide information about the file size and the time that may be required to download it before users commit to downloading the entire file.

• Careful use of graphics. Image formats should be chosen carefully. Most images on the WWW are represented using either the GIF or JPEG formats. In general, the GIF format works best for cartoon-like images, while the JPEG format works better for real-life images. Both formats provide mechanisms for reducing the size of the resulting file. Options include reducing the size of the palette in a GIF file and varying the level of compression used when creating a JPEG file. Careful use of these capabilities can significantly reduce file sizes without losing any content. Consider the benefits and drawbacks before using animated graphics, reuse graphics to reduce download times, and replace graphics with text if this can be done without losing information. Providing low-resolution versions of graphics that can be downloaded and displayed quickly is also important, as is considering the possibility that users with slow connections will simply disable image loading.

• Use the features of HTML. Often, graphics are used where built-in features of HTML could accomplish the same goal. For example, it is less efficient to use small graphics for bullets to format lists or to create horizontal lines. Retrieving even the smallest file requires that a network connection be established (unless the keep-alive capability is being used) and header information be transmitted. In all but the most unusual of situations, downloading a graphic will require longer than downloading one or two HTML tags.

• Generate clean HTML. Eliminating unnecessary characters from HTML source code can reduce the size of a file by 10% or more. This is even more important when using any of the standard HTML authoring programs because extra tabs and line breaks are inserted to make the HTML source easier to read. These authoring tools also insert a variety of unnecessary tags (e.g., identifying the name of the program used to generate the HTML).

• Minimize delays between download and display. HTML is rendered starting at the top of the file. Whenever an object (e.g., an image) is encountered, its size must be determined before the information can be rendered correctly. Although various strategies are employed to reduce this problem (e.g., displaying small icons and rerendering the document when the size of the image is determined), defining the height and width of every object can help eliminate this problem. Placing graphics lower on the page may also reduce the problem users experience due to the delays associated with downloading images. If appropriate, scripts can be used to preload graphics. Tables must be used carefully because they can result in longer delays before information is displayed for the user, and frames are even more problematic because multiple documents must be downloaded before any information is made available to the user.

• Consider the time required to establish connections and download information. Unless the stay-alive feature is used, each document will require a new connection between the user's computer and the server. Even with stay-alive, each document results in extra header information being transmitted. As a result, there are benefits to reducing the number of documents that are downloaded. At the same time, there are situations where downloading multiple smaller documents may prove faster than downloading one large document. A balance must be reached between the time required to establish new connections and the time required to download information. If establishing connections is a slow process, downloading one larger file that provides all of the information a user may need may be more appropriate. If new connections can be established quickly, downloading multiple small files may prove effective.

• Consider parallelism. While this is not always a good option, there are times when dividing a file into several smaller files can prove beneficial. This can help if the browser downloads multiple documents in parallel. Once the main HTML document is downloaded, it is common for several supporting subdocuments (as many as four or more) to be downloaded at the same time. Dividing a document into several smaller documents can, under the right conditions, reduce the delays users experience. To illustrate this concept, a 4K image and a 128K image were each divided into four files of approximately the same size. The original images, as well as the divided images, were stored on several servers across the United States. The images were downloaded repeatedly over a T-1 line, and the delays were recorded. Dividing the 4K image into four parts saved an average of 200 msec, while dividing the 128K image saved over 750 msec. Of course, the exact benefits will depend on a number of factors, including the number of subdocuments referenced in the main HTML document, the network conditions when the documents are retrieved, and the speed of the user's Internet connection. A sufficiently slow Internet connection can prove to be the limiting factor when trying to reduce download delays. If this is the case, the technique described above is unlikely to provide any benefit.

Evaluating Network Delays

Iterative design is critical for success, and WWW development happens at a rapid pace. Given the potential for network delays to negatively impact user experiences and perceptions, material developed for distribution via the WWW should be evaluated in the context of realistic delays. Five criteria should be considered when selecting a technique that allows documents to be viewed (Sears & Jacko, 2000):

• *Setup time:* Minimize the time and effort required to prepare when evaluating a new collection of materials.

• *Reliability:* Techniques that require the use of an Internet service provider have a greater risk of failure than techniques that do not. Such failures may be realistic but do not contribute to the process of evaluating the appropriateness of the delays users experience.

• *Accuracy of the experience:* The technique must allow designers, users, or study participants to experience realistic delays as documents are viewed. Delays due to access speeds (i.e., Internet connection) as well as the Internet itself should be experienced.

• *Replication:* The ability to replicate network conditions will allow testing to be distributed across several days and reproduced later if necessary.

• *Flexibility:* It should be possible to experience multiple network conditions on demand. For example, it can be useful to experience delays that are representative of 1 p.m. on a weekday at one moment and then switch to delays representative of 1 a.m. on a weekend at another. Flexibility also refers to the ability to fine-tune the experience. Although delays may be highly variable, it may prove useful if the documents can be viewed while experiencing delays that are representative of relatively good, average, or relatively poor network conditions.

Organizations employ a variety of techniques for evaluating the materials they develop for the WWW. Three approaches are described below with advantages and disadvantages listed for each (Sears & Jacko, 2000).

Using the Actual Network to Manipulate Delays

In some situations, evaluations can be conducted using the same network that the end users will ultimately use. Materials can be placed on a computer in one location and accessed from a computer in another. This provides a realistic experience but may require significant setup if numerous documents must be copied to the remote computer. This arrangement can also be reversed, with users logging into a remote computer and accessing materials stored locally. This will reduce the setup time, but also reduces the accuracy of the experience because all information now makes two trips across the network instead of one. Because both approaches use the actual network, neither can allow users to experience different network conditions on demand, and network conditions may be difficult to reproduce.

Using a Local Computer and a Network Connection to Manipulate Delays

A second approach involves storing documents on a local computer. The documents are then accessed using a second local computer (or in some instances the same computer) via some kind of network connection. The simplest alternative would be to access the company's LAN via a modem. This requires minimal setup and introduces realistic access speeds, but it fails to incorporate the potentially extensive delays associated with Internet usage. As a result, the bandwidth users experience may be realistic, but the latencies are not.

Using the same arrangement, but accessing the materials via an Internet service provider (ISP) results in more realistic latencies, but how realistic they are depends on the location of the ISP relative to the company's connection to the Internet. If a local ISP is used, latencies may be unrealistically short. Using a remote ISP should provide more realistic delays but will also result in higher costs and more opportunities for technological failures.

The local ISP option is similar to yet another common technique of having employees download documents from their home computers. With each of these approaches, the experience depends on the current network conditions, flexibility is reduced, the ability to replicate the study is reduced, and reliability can be a concern.

Using Network Simulation Software to Manipulate Delays

Simulating network delays can require minimal setup, provide realistic experiences, allow numerous network conditions to be experienced on demand, and minimize the possibility of technological failure while allowing network conditions to be replicated on demand. In addition, the documents can be developed, stored, and viewed using a single computer. Refer to Sears (in press) for an example of a collection of tools that allow WWW materials to be evaluated using simulated network delays.

Designing for Small Screens

It is becoming increasingly common to access WWW-based content using devices that provide small displays (or larger displays with limited resolution). Wireless PDAs, mobile phones, handheld PCs, Web TV, and various other devices now provide Internet access. If individuals will be accessing materials using devices other than a standard desktop and laptop computer, small displays should be considered.

Several articles discuss the issues involved in designing for small displays (e.g., Buyukkokten, Garcia-Molina, Paepcke, & Winograd, 2000; Buyukkokten, Garcia-Molina, & Paepcke, 2001; Jones, Marsden, Mohd-Nasir, Boone, & Buchanan, 1999; Kamba, Elson, Harpold, Stamper, & Sukaviriya, 1996). As would be expected, users tend to be less efficient, and alternative interaction mechanisms are suggested to allow for more efficient navigation. A great deal of additional research and development effort is currently focused on these issues. The following guidelines are important:

• The design should be simplified. Clutter should be minimized, unnecessary information eliminated, and writing reviewed to ensure that it is concise yet accurate. Use of features that consume larger quantities of screen space or rely on high-resolution displays should be minimized, including graphics, tables, and frames. If graphics are used, remember that users may turn image loading off.
• Given limited screen space, information organization becomes even more important. The information users are likely to need should be placed such that it will be available without scrolling. If scrolling is required, it should be limited to one direction (vertical is better than horizontal) because forcing users to scroll both vertically and horizontally to view a document is likely to result in frustration. An alternative is to divide information into several smaller documents and provide an index. Effective navigational support is critical.
• When retrieving material from the WWW using a small device, individuals are likely to have specific goals. By understanding the user's goals, the specific information that will support these goals can be delivered and formatted more effectively.

CONCLUSIONS

Universal usability is a concept that is still evolving. There are legal and economic issues, and additional user groups and technological issues could also be discussed. In short, it is not possible to provide a detailed discussion that covers all of the issues associated with universal usability in a chapter of this length. Therefore, we end this chapter by providing pointers to a variety of additional resources, including books, conferences, and WWW sites.

Various books either focus directly on the issues associated with universal usability or include chapters that discuss relevant issues. *Extra-Ordinary Human–Computer Interaction* (Edwards, 1995b) is an edited collection of 19 chapters that cover the fundamentals, several case studies, overviews of research, and examples of practical applications of these ideas. *User Interfaces for All* (Stephanidis, 2001) is an edited collection of 30 chapters that discuss the underlying concepts as well as methods and tools for supporting universal usability. *The Human–Computer Interaction Handbook* (Jacko & Sears, 2003) contains numerous chapters that are relevant when designing for diverse users.

A growing number of conferences include useful articles, such as the Assets conferences organized by SIGCAPH (SIGCAPH, 2001), CUU conferences organized by SIGCHI (SIGCHI, 2001), RESNA conferences organized by the Rehabilitation Engineering and Assistive Technology Society of North America (RESNA, 2001), CSUN conferences organized by the Center on Disabilities at California State University, Northridge (California State University, Northridge, 2001), and the UAHCI conferences (UAHCI, 2001).

Finally, there are numerous Web sites that contain relevant information. Given the volume of Web sites and how rapidly they change, a limited number of stable sites that provide useful information and pointers to additional sites are included here. The Trace Center provides a variety of documents pointers to additional resources on its Web site (Trace Center, 2001), as do the Universal Usability (Universal Usability, 2001) and Falling Through the Net (Digital Divide, 2001) Web sites. The World Wide Web Consortium's site contains valuable information, especially in the section describing the Web Accessibility Initiative (W3C, 2001). Perhaps the most useful information contained at these three Web sites are the numerous pointers they provide to many additional resources.

ACKNOWLEDGMENTS

This material is based on work supported by the National Science Foundation (NSF) under grants IIS-9910607 and IIS-0121570. Any opinions, findings, and conclusions or recommendations expressed in this material are those of the author and do not necessarily reflect the views of the National Science Foundation.

REFERENCES

Bederson, B. B., Hollan, J. D., Druin, A., Stewart, J., Rogers, D., & Proft, D. (1996). *Local tools: An alternative to tool palettes. Proceedings of user interface and software technology* (pp. 169–170). New York: Association for Computing and Machinery.

Belge, M. (1995, January). The next step in software internationalization. *Interactions, 2*(1), 21–25.

Bernard, M., Mills, M., Frank, T., & McKnown, J. (2001). Which fonts do children prefer to read online? [online]. *Usability News: A Newsletter of the Software Usability Research Laboratory.* Available: http://wsupsy.psy.twsu.edu/surl/usabilitynews/3S/font.htm (Last visited, November 6, 2001).

Bruckman, A., & Bandlow, A. (2003). HCI for kids. In J. Jacko & A. Sears (Eds.), *The human–computer interaction handbook.* Mahwah, NJ: Lawrence Erlbaum Associates.

Buyukkokten, O., Garcia-Molina, H., Paepcke, A., & Winograd, T. (2000). Power browser: Efficient Web browsing for PDAs. In T. Turner, G. Szwillis, M. Czerwinski, F. Paterno, & S. Pemperton (Eds.), *Proceedings of CHI 2000* (pp. 430–437). New York: Association for Computing and Machinery.

Buyukkokten, O., Garcia-Molina, H., & Paepcke, A. (2001). Accordion summarization for end-game browsing on PDAs and cellular phones. In *Proceedings of CHI 2001* (pp. 213–220). New York: ACM.

CSUN WWW site. (2001). California State University Northridge Center on Disabilities [online]. Available: http://www.csun.edu/cod/ (Last visited, November 6, 2001).

Carmichael, A. R. (1999). *Style guide for the design of interactive television services for elderly viewers.* Winchester, New York: Independent Television Commission.

Casali, S. P. (1992). Cursor control device used by persons with physical disabilities: Implications for hardware and software design. In *Proceedings of the Human Factors Society 36th Annual Meeting* (pp. 311–315). Santa Monica, CA: Human Factors and Ergonomics Society.

Charness, N., Bosman, E. A., & Elliot, R. G. (1995). *Senior-friendly input devices: Is the pen mightier than the mouse?* Paper presented at the 103rd annual convention of the American Psychological Association, New York.

Children's Online Privacy Protection Act. (1998). Available: http://www.ftc.gov/ogc/coppa1.htm (Last visited: February 19, 2002).

Czaja, S. J., & Lee, C. C. (2003). Designing computer systems for older adults. In J. Jacko & A. Sears (Eds.), *The human–computer interaction handbook.* Mahwah, NJ: Lawrence Erlbaum Associates.

Digital Divide. (2001). Falling through the Net [online]. Available: http://www.digitaldivide.gov (Last visited, November 6, 2001).

Dix, A. (2003). Network-based interaction. In J. Jacko & A. Sears (Eds.), *The human–computer interaction handbook.* Mahwah, NJ: Lawrence Erlbaum Associates.

Druin, A. (1998). *The design of children's technology.* San Francisco: Kaufmann.

Edwards, A. (1995a). Computers and people with disabilities. In A. Edwards (Ed.), *Extra-ordinary human–computer interaction* (pp. 19–43). New York: Cambridge University Press.

Edwards, A. D. N. (1995b). *Extra-ordinary human–computer interaction: Interfaces for users with disabilities.* Cambridge, England: Cambridge University Press.

Inkpen, K. (2001). Drag-and-drop versus point-and-click: Mouse interaction styles for children. *ACM Transactions Computer–Human Interaction, 8*(1), 1–33.

Jacko, J., & Sears, A. (2003). *The human–computer interaction handbook.* Mahwah, NJ: Lawrence Erlbaum Associates.

Jacko, J. A., Sears, A., & Borella, M. S. (2000). The effect of network delay and media on user perceptions of web resources. *Behaviour and Information Technology, 19,* 427–439.

Jacko, J., Vitense, H., & Scott, I. (2003). Perceptual impairments. In J. Jacko & A. Sears (Eds.), *The human–computer interaction handbook.* Mahwah, NJ: Lawrence Erlbaum Associates.

Joiner, R., Messer, D., Light, P., & Littleton, K. (1998). It is best to point for young children: A comparison of children's pointing and dragging. *Computers in Human Behavior, 14*(3), 513–529.

Jones, M., Marsden, G., Mohd-Nasir, N., Boone, K., & Buchanan, G. (1999, May). *Improving Web interaction on small displays.* Paper presented at WWW Conference, Toronto, CA.

Kamba, T., Elson, S. A., Harpold, T., Stamper, T., & Sukaviriya, P. (1996). Using small screen space more efficiently. In M. Tavber, V. Bellotis, R. Jeffries, J. MacKinlay, & J. Nielsen (Eds.), *Proceedings of CHI 1996* (pp. 383–390). New York: Association for Computing and Machinery.

Keates, S., Clarkson, J., & Robinson, P. (2000). Investigating the application of user models for motion-impaired users. In *Proceedings of ASSETS 2000* (pp. 129–136). New York: Association for Computing and Machinery.

Kline, D. W., & Schieber, F. (1985). Vision and aging. In J. E. Birren & K. W. Schaie (Eds.), *Handbook of the psychology of aging* (pp. 296–331). New York: Von Nostrand Reinhold.

Kraut, J. A., & McCabe, C. P. (1994). The problem of low vision. In D. M. Albert, F. A. Jakobiec, & N. L. Robinson (Eds.), *Principles and practices of ophthalmology* (pp. 3664–3683). W. B. Saunders Company.

Luong, T., Lok, J., Taylor, D., & Driscoll, K. (1995). *Internationalization: Developing software for global markets.* New York: Wiley.

Marcus, A. (2003). Global and intercultural user-interface design. In J. Jacko & A. Sears (Eds.), *The human–computer interaction handbook.* Mahwah, NJ: Lawrence Erlbaum Associates.

Newell, A. (1995). Extra-ordinary human–computer interaction. In A. D. N. Edwards (Ed.), *Extra-ordinary human–computer interaction: Interfaces for users with disabilities* (pp. 1–3). Cambridge, England: Cambridge University Press.

Newell, A., Carmichael, A., Gregor, P., & Alm, N. (2003). Information technology for cognitive support. In J. Jacko & A. Sears (Eds.), *The human-computer interaction handbook.* Mahwah, NJ: Lawrence Erlbaum Associates.

O'Donnell, S. (1994). *Programming for the world: A guide to internationalization.* Engelwood Cliffs, NJ: Prentice-Hall.

Park, D. (1992). Applied cognitive aging research. In F. I. M. Craik & T. A. Salthouse (Eds.), *The handbook of aging and cognition* (pp. 44–494). Hillsdale, NJ: Lawrence Erlbaum Associates.

Radwin, R. G., Vanderheiden, G. C., & Lin, M. L. (1990). A method for evaluating head-controlled computer input devices using Fitts' law. *Human Factors, 32*(4), 423–438.

Ramsay, J., Barbesi, A., & Preece, J. (1998). A psychological investigation of long retrieval times on the World Wide Web. *Interacting With Computers, 10*, 77–86.

RESNA. (2001). Rehabilitation Engineering and Assistive Technology Society of North America WWW site [online]. Available: http://www.resna.org/ (Last visited, November 6, 2001).

Riviere, C. N., & Thakor, N. V. (1996). Effects of age and disability on tracking tasks with a computer mouse: Accuracy and linearity. *Rehabilitation Research and Development, 33*(1), 6–15.

Rogers, W., & Fisk, A. (2000). Human factors, applied cognition, and aging. In F. I. M. Craik & T. A. Salthouse (Eds.), *The handbook of aging and cognition.* Mahwah, NJ: Lawrence Erlbaum Associates.

Russo, P., & Boor, S. (1993). How fluent is your interface? Designing for international users. In S. Ashland, K. Mullet, A. Henderson, E. Hollnagel, & T. White (Eds.), *Proceedings of INTERCHI 1993* (pp. 342–347), New York: ACM.

Salomon, G. (1990). New uses for color. In B. Laurel (Ed.), *The art of human–computer interface design* (pp. 269–278). Reading, MA: Addison-Wesley.

Schieber, F., Fozard, J. L., Gordon-Salant, S., & Weiffenbach, J. W. (1991). Optimizing sensation and perception in older adults. *International Journal of Industrial Ergonomics, 7*, 133–162.

Sears, A. (in press). Simulating network delays: Applications, algorithms, and tools. *International Journal of Human–Computer Interaction.*

Sears, A., & Jacko, J. A. (2000). Understanding the relationship between network quality of service and the usability of distributed multimedia documents. *Human Computer Interaction, 15*, 43–68.

Sears, A., Jacko, J. A., & Dubach, E. M. (2000). International aspects of WWW usability and the role of high-end graphical enhancements. *International Journal of Human–Computer Interaction, 12*(2), 243–263.

Sears, A., Karat, C.-M., Oseitutu, K., Karimullah, A., & Feng, J. (2001). Productivity, satisfaction, and interaction strategies of individual with spinal cord injuries and traditional users interacting with speech recognition software. *Universal Access in the Information Society, 1*, 4–15.

Sears, A., & Young, M. (2003). Physical disabilities and computing technologies: An analysis of impairments. In J. Jacko & A. Sears (Eds.), *The human–computer interaction handbook*. Mahwah, NJ: Lawrence Erlbaum Associates.

U.S. General Services Administration. (2002). Available: http://www.section508.gov/ (Last visited, February 19, 2002).

Shneiderman, B. (2000). Universal usability: Pushing human–computer interaction research to empower every citizen, *Communications of the ACM, 43*(5), 85–91.

SIGCAPH. (2001). *ACM Special Interest Group on Computers and the Physically Handicapped* [online]. Available: http://www.acm.org/sigs/sigcaph/ (Last visited, November 6, 2001).

SIGCHI. (2001). *ACM Special Interest Group on Computer–Human Interaction* [online]. Available: http://www.acm.org/sigs/sigchi/ (Last visited, November 6, 2001).

Smith, N. W., Sharit, J., & Czaja, S. J. (1999). Aging, motor control, and performance of computer mouse tasks. *Human Factors, 41*(3), 389–396.

Stephanidis, C. (2001). *User interfaces for all*. Mahwah, NJ: Lawrence Erlbaum Associates.

Trace Center WWW site. (2001). Available: http://www.trace.wisc.edu (Last visited, November 6, 2001).

Trewin, S., & Pain, H. (1999). Keyboard and mouse errors due to motor disabilities. *International Journal of Human–Computer Studies, 50*, 109–144.

UAHCI. (2001). Universal access in human–computer interaction [online]. Available: http://uahci.ics.forth.gr/ (Last visited, November 6, 2001).

Universal Usability WWW site. (2001). Available: http://www.universalusability.org (Last visited, November 6, 2001).

W3C WWW site. (2001). World-Wide Web Consortium [online]. Available: http://www.w3.org (Last visited, November 6, 2001).

Wood, J. M., & Troutbeck, R. J. (1994, October). Effect of age and visual impairment on driving and vision performance. *Transportation Research Record, 1438*, 84–90.

WHO. (2000). *International classification of functioning, disability, and health*. Geneva, Switzerland: Author.

3

A Cultural Comparison of Website Design From a Usability Engineering Perspective

Masaaki Kurosu

National Institute of Multimedia Education

INTRODUCTION

As the Internet has grown globally, culture is now a crucial issue for Website design. Many Websites, including search engines and shopping and corporate sites, are now targeting both domestic and international audiences. Although English is the predominant language for global Websites, many savvy Web designers have realized that layout, information structure, and content needs to accommodate cultural and regional preferences. Such design preferences are necessary to consider in the context of usability, effectiveness, efficiency, and high user satisfaction, and to meet ISO 13407 (1999) standards to achieve human-centered design. The cultural aspects of Website design are the focus of this chapter, along with how usability design guidelines can help designers create more effective and usable sites for commercial purposes.

Research on cultural problems is usually based on ethnographic field study methods that include interviewing and observation. This general overview of the problem is useful for clarifying which cultural aspects of Website design should be targeted.

With field research methods, the internal processes of each individual user are examined, that is, the user's motivation, the "why" and "how" the user needs specific information, and which aspect of the Website was effective or ineffective relative to the user's specific culture.

This chapter provides an overview and review of two important cultural dimensions that are important to Website design. Then, certain cultural aspects of Website design are compared based on localization and globalization. Cultural aspects of Websites are then considered by analyzing the Website's usability from a psychological point of view.

This chapter presents a high-level cultural analysis on Website design. It is expected that this will serve to open the door for future research. The use of field research methodology is strongly recommended for scrutinizing cultural matters relating to Website design. Future research on the cultural aspects of Website design should incorporate field work methods in order to clarify what kind of information people from each cultural region expect and how it should be designed.

DIMENSIONS OF CULTURE

There are two dimensions of culture, one concerning cultural variety and the other, its depth. *Cultural variety* is a qualitative dimension that ranges from global to individual culture. *Depth of culture* is a quantitative dimension that ranges from a shallow to a deep level. Using these two dimensions, targeted aspects of culture are explored in this chapter.

There are many definitions of culture in the history of cultural anthropology, including Tylor's (1958). When we discuss culture in terms of Website design, we define it from a functional viewpoint. Thus, Suzuki's (1997) definition is relevant here, that "the culture is the response pattern shared by some specific group of people that is shaped through interaction with the environment." In the context of the Website, "the response pattern" is *how* people will interact with the Website, and "interaction with the environment" is the interaction with the Website through the PC environment, including the browser.

The first dimension of culture is its variety. Regarding the "specific group of people," there are various groups with regard to the interaction. In Fig. 3.1, I describe whole varieties of culture (Kurosu, 2001). *Global culture* is what is common to all human beings. *Country culture* and *ethnic culture* are sometimes confusing to delineate. American culture and Chinese culture are country cultures, for example, but the United States includes many ethnic groups, such as the Chinese, who have their own ethnic culture common to those living in the People's Republic of China, Singapore, and other countries. What is important here is that because Chinese people living in the United States are acclimated to American culture, whose group's culture differs from the culture of Chinese living elsewhere. Hence, we differentiate country culture and ethnic culture when we look at an ethnic group living in a specific country. Because it is difficult to clearly differentiate these two concepts, I combine these two concepts depending on the

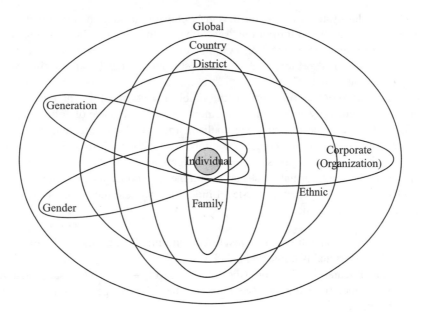

FIG. 3.1. Diagram of dimensions of culture.

context in this work. That is, I use the word *culture* to mean country culture for the most part when describing corporate Websites, because corporate activity is, in most cases, specific to that country. The word *culture* is also used to describe ethnic culture, because the way people think can be strongly affected by ethnic culture.

Generation culture explains the difference between behavioral patterns of adults and youth. *District culture* explains the minor differences in patterns within a specific country. Although *gender culture* is strongly affected by country culture and ethnic culture, it is also an independent and unique aspect of culture. Further, although the *corporate culture* is, in most cases, dependent on country culture and could be considered in the context of country culture, we also distinguish international industry and local industry. At another level, there is the unique culture of the family. And, at the most basic level, each individual has his or her behavioral and mental patterns. All these dimensions of culture play important roles in everyday behavior. The discussion in this chapter will focus solely on country culture and ethnic culture.

Another type of cultural dimension is its depth. Hofstede (1991) proposed a famous onion diagram with layers labeled as cultural values, rituals, heroes, symbols, and practices. In his diagrammatic model, symbols are located at the surface and values at its core. A similar concept is expressed as an "island model" by Carter (1997), in which the top level implies architecture, clothing, and so forth, and is visible over the sea-surface. The second level, which implies ways of thinking, value systems, and so forth, is just below the surface and is invisible. And the third level, at the bottom of the "sea," implies fundamental human behaviors, such as eating and communication, common to all human beings. What is common to these models is that there is always at least a surface level that is visible and a deep level that is invisible.

In the context of the user interface, the visible aspects of culture include date and time format, currency, units of measurement, symbols, and use of color. Invisible aspects include value systems, nationalism, the social context, and aesthetics. These distinctions can, of course, be applied to Website design. So we should consider both visible aspects, including widgets and their layout, and invisible aspects, including information structure and the core concept of the site.

Shade (1998) proposed a four-level model to evaluate the localization of the user interface, including

1. Minimal localization (the lowest level)—minimal requirements, for example, when a manual is translated into Japanese.
2. Fundamental localization (the second level)—when necessary changes are made, for example, the label of the GUI widget is translated into Japanese.
3. Adequate level localization (the third level)—when necessary elements are refined, for example, visual design is adapted for the Japanese user.
4. The high-level user experience (the top level)—when the application is totally adapted for the user, for example, the help system is rewritten to suit the emotional aspects of the user in addition to giving necessary information.

Although the direction is atypical compared to the usual model, it proposes what should be done for each level of depth.

One final point regarding culture is its interaction aspect. Interaction with the Website can be viewed as a communication process. In this respect, the model of interpersonal communication proposed by Hoshino (1989) is used as a framework for understanding interaction behavior on a Website.

Website browsing is one entity for the user, and the other is the interaction of the Website within the browser. For example, the Website sends the message through visual and auditory channels to the user. The user decodes the information and interprets it, then starts the next action based on his needs and purpose for browsing the information accumulated to that point. The Website receives the user's action as a message, usually in the form of mouse clicks or text input. Then the system returns a response, for example, the next page, as defined by the program. In this framework, there are two cultures, that is, the culture on the part of the user and on the part of the Website. The interpretation of the Website information can be misunderstood when the user input triggers an unintended action on the Web page. When Web design is based on cultural diversity, this difference is minimized as much as possible.

WEBSITE DESIGN AND CULTURE

Today, Websites are designed in developed and in developing countries. A Website that targets a specific country should follow the cultural rules and customs explicitly or implicitly possessed by people of that country. Because Web designers living in that country will consciously or unconsciously be aware of those rules and customs, they will naturally design Websites in accordance with their culture. A challenge arises when people from other countries create Websites for other cultures with which they are not familiar. In such a case, it is recommended that they contact a local designer or a usability professional working in that country and ask for some assistance in designing and testing the design with local users. For example, international companies frequently face this situation when they try to implement company Websites outside their native home office. A simple translation of the original Website designed in one country will often be less than successful because of discrepancies between the Website design and the culture of the targeted country. This is what Shade (1998) described as the lowest level localization.

Such a situation is illustrated in Fig. 3.2, with different types of design strategies. Two distributions in this figure represent two different cultures. Basing the design of a Website on one culture for use in two countries will inevitably result in an ethnocentric design. The ethnocentric design, intentionally or unintentionally, suggests a cultural hegemony. For example, designing a Website using Fahrenheit degrees, inches, and pounds will be, at the very least, puzzling or unintelligible to many Japanese users. Ethnocentric sites are frequently viewed as unattractive, disappointing, and sometimes even insulting to a targeted user.

Other design strategies are based on a consideration of more than one culture. However, integrating one extreme characteristic from a targeted culture may result

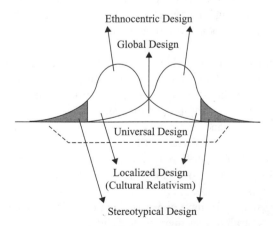

FIG. 3.2. Illustration of universal design strategy for two different cultures.

in a stereotypical design. For example, in Japan, the combination of red and white colors is frequently used in happy celebrations, such as opening ceremonies. But a Website overwhelmed with this color combination can result in a site that could be considered contrived and/or unattractive to Japanese users.

The more practical approach is to utilize a few typical aspects of each culture. This approach is integral to this author's stance in this chapter, is one of localized design, or *localization*, based on cultural relativism, and also is the one that is commonly accepted by diverse targeted audiences.

Taking the logical boolean "AND" operant of two distributions leads to a global design, or *globalization*. This design approach is based on common characteristics among different cultures from the viewpoint of human beings. Taking the logical boolean "OR" operant of them means a *universal design*, or one-design-for-all approach. This approach considers the diverse variations among the audiences and, hence, the infinite variations in the design. Though the goal of this approach seems reasonable, its implementation is quite difficult to achieve.

CULTURAL ASPECTS OF WEB USABILITY INTERFACE DESIGN GUIDELINES

As stated in the introduction, usability became recognized as one critical aspect of Website design, as it became clear that Websites could become even more effective for commercial purposes when the audience is clearly defined. Since the mid-1970s, many Website usability guidelines have been proposed, as shown in Table 3.1. Therefore, this section focuses on the usability aspects of Website design from a cultural perspective.

Tables 3.1 and 3.2 summarize the publication dates of Web usability guidelines chronologically, including books and Websites in both the United States and Japan. Table 3.3 lists online global design references on Web usability and cultural diversity.

As seen in Table 3.3, there is a 2- to 3-year delay between the United States and Japan, with Japan lagging behind. The reason for this delay might be the delay of the launch of usability engineering activity in Japan. Unlike in the United States, people in industry in Japan, especially project managers, were not aware of usability, or "quality in use," issues and instead focused on functionality, performance, and reliability. The importance of usability has come to the forefront in Japan with the advent of ISO 13407 (1999) and publications on Web usability in the United States. Since 1999, three famous guidelines published in the United States have been translated into Japanese. As a result of these translations, many guidelines written by Japanese authors have appeared, but none of them have been translated into English. It is strange that there is no English translation of such Japanese guidelines, because these guidelines, originated in Japan, might be effective and valuable for localizing U.S. Websites in Japan. It seems logical to believe that there

TABLE 3.1
References on Web Usability

	Web Usability	
Year	United States	Japan
1997	J. M. Spool et al.	H. Sasaki
1998	J. Fleming C. S. Blyth C. Forsythe et al.	
1999	J. M. Spool et al. P. J. Lynch & S. Horton C. Kilian	M. Kurosu et al.

2000	J. Nielsen
	S. Krug
	
	M. Pearrow
		J. Nielsen, Trans. by T. Shinohara
		J. M. Spool et al., Trans. by
		   T. Shinohara
		H. Uchida,
		   Y. Yamamoto, &
		   J. Y. Tajima

2001	L. Lopuck T. Brinck et al. N. Campbell E. Dustin et al. J. Nielsen & M. Tahir,	Bebit M. Pearrow, Trans. by K. Modeki S. Krug, Trans. by E. Nakano Cis-z.
2002*1	A. Badre, A. N. Badre & M. Pearrow	

is a need to introduce Japanese usability guidelines into English. This author's hypothesis is that because U.S. Website usability guidelines are predominant they are likely culturally localized for North American audiences.

My concern here is to what extent Website usability guidelines are affected by cultural factors, even though Japan now has access to almost all of the relevant information from the United States. The question is how Website design should be approached differently from culture to culture.

A CULTURAL COMPARISON OF FIFTEEN U.S. AND JAPANESE WEBSITE DESIGNS

Because it is not sufficient to investigate the relationship between Website usability and cultural factors by considering only certain guideline items, I propose another approach to compare Website designs (Kurosu, Motoki, & Itoh, 2001; Motoki, 2001).

The purpose of this site comparison was to find out if there were actually differences in terms of usability among Websites that were designed in Japan

TABLE 3.2
Online References on Web Usability (For URLs, refer Appendix 1)

| Year | Web Usability | |
	United States	Japan
1995	(7)	
1996	(6)	(1)
1997	(22)	(8)
	(11)	
1998	(10)	(13)
	(16)	(17)
	(21)	
1999	(19)	
	(15)	
	(5)	
	(18)	
2000	*2	(14)
		(20)
2001	(4)	(12)
	*2	(3)
		(2)
		(9)
2002*1	*2	

TABLE 3.3
Global Design References on Web Usability and Cultural Diversity

| Year | Global Design | |
	United States	Japan
1992	D. Taylor	
1993		
1994		
1995	T. Fernandes	
1996	J. Nielsen & E. M. Del Galdo (Eds.)	
1997		
1998		
1999		
2000	B. Esselink	

compared with those designed in the United States. A total of 30 Websites were compared, including 10 search engines, 10 shopping sites, and 10 corporate sites, of which half were designed in Japan and half in other countries. Primarily U.S. domestic and international sites were selected in all three Website categories.

In order to apply a checklist rating method, items were extracted from Nielsen (2000a) and Spool (2000). The list includes categories such as page design, fonts, and navigation methods, and a total of 118 items were checked. A 3-point rating scale was used to evaluate the degree of conformity to the guidelines.

The average rating score of 118 usability design guideline items was then calculated. Surprisingly, the research findings sites follow a sequential order for Japanese Websites and U.S. Web. Websites in the United States were not typically rated higher than Japanese Websites, or vice versa. For the purpose of confirming these scores, an analysis of variance (ANOVA) was applied to the data. It was applied to the search engines, shopping sites, and corporate sites, respectively, for a total of 30 sites. ANOVA results showed no significant difference between the Japanese and the U.S. Websites at a 5% level. From a statistical point of view, even the lack of a significant difference does not imply that the variables are from an identical population or not from different populations. This finding merely reflects the fact that Japanese Websites and U.S. Websites are indistinguishable in terms of usability.

These results correlate with the fact that many usability guidelines published in the United States have been translated into Japanese and, thus, many Japanese Website designers follow these design guidelines. It is true that the Japanese tend to trail the United States in other areas, like politics, art, and technology. The result of this experiment can be interpreted in terms of the surface level of cultural aspects by such a phenomenon like other areas of Japanese culture.

CONCLUSION

There are many guidelines on international interface design and important tips for localization (see Appendix 1).

In the previous sections, we confirmed that American and Japanese Website design usability shares many similar traits at the surface level. But in order for Website design to be adapted to a targeted culture, it is necessary to factor in localization issues at the shallow level of culture. For example, Website designers need to omit MM/DD/YY or DD/MM/YY date formats on Japanese Websites and use YY/MM/DD format instead.

In addition to these shallow aspects, designers need to consider the deeper aspect—The information content and the core concept of a Website—more fully because it is deeply linked to way of life, and job procedures in a specific culture needs to be thoroughly comprehended.

The human-centered design approach in ISO 13407 (1999) emphasizes the importance of specifying "context of use" as the primary design process, as shown in Fig. 3.3. Specifying context of use includes understanding how people live and do their jobs by collecting information through interviews, observation, and other techniques. This legislation is directly related to the understanding of deep levels of culture of the targeted audience. In addition, the process of evaluating designs against user requirements is also central to the purpose of confirming that Website

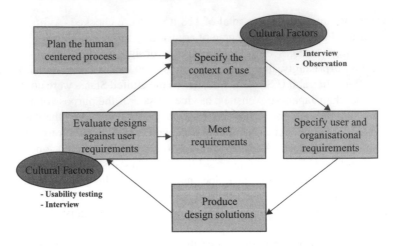

FIG. 3.3. Human-centered design process for interactive systems in ISO 13407.

design is adapted to the targeted culture. Putting an emphasis on these two design requirements is key to making global Web designs.

Based on the analysis of Web design usability guidelines and cultural comparative research, this author concluded that Website usability is universal at the surface level of culture. As long as Websites are designed to take into account local differences such as measurement units, they will probably be accepted by non-native users. Because Website usability guidelines proposed to date have concentrated on surface-level features of design (i.e., visual aspects of usability), that cultural dissonance occurs. Website designers need to consider usability that closely relates to deeper cultural issues. For example, because the Japanese celebrate the coming of the new year much the same as Americans do Christmas, specially priced bargains on the first three days of the new year are an attractive feature for shopping sites. Thus, it is necessary for Website designers to make a greater effort in obtaining a context of use of information by applying field study techniques to understand customs and habits of the people living in targeted cultures by using ethnographical references. It is not enough for designers to merely follow Website usability guidelines if they wish to make sites sufficiently effective, usable, and satisfying to global audiences. My hope is that this chapter provides evidence of an immediate call to action to open doors for future ethnographic research.

APPENDIX 1

Web sites listed in Table 3.2 are shown here in alphabetical order.

1. @laplace-lab (Note: In Japanese)
   (http://www.laplace-lab.org/index.shtml)

2. All About Japan (Note: In Japanese)
   (http://allabout.co.jp/)
3. Catchball 21 Internet Consulting (Note: In Japanese)
   (http://www.cb21.co.jp/)
4. E-Commerce User Experience (Note: Link to purchase 2001 guidelines)
   (http://www.nngroup.com/reports/ecommerce/)
5. http://builder.cnet.com/webbuilding/pages/Graphics/CTips2/ By Matt
   Rosoff (4/28/98) entitled More Great Tips from CNET Designers, Part 2:
   Designing Pages and Sites
6. Guidelines for Designing Usable World Wide Web Pages, *José A. Borges,
   Israel Morales, & Néstor J. Rodríguez* (http://www.acm.org/sigchi/chi96/
   proceedings/shortpap/Rodriguez/rn _txt.htm)
7. Guidelines for Multimedia on the Web, December 1995, Jakob Nielsen
   (http://www.useit.com/alertbox/9512.html)
8. HotWired Japan (Note: In Japanese)
   (http://www.hotwired.co.jp/webmonkey/)
9. IBM Developer Works (Note: In Japanese)
   (http://www-6.ibm.com/jp/developerworks/)
10. IBM Ease of Use Web Guidelines (http://www-3.ibm.com/ibm/easy/eou_ext.
    nsf/publish/572)
11. Improving Web Site Usability and Appeal Link broken (http://msdn.
    microsoft.com/library/default.asp?url=/library/en-us/dnsiteplan/html/
    IMPROVINGSITEUSA.asp)
12. japan.internet.com (Note: In Japanese)
    (http://www.japan.internet.com/)
13. KeiYu HelplLab (Note: In Japanese)
    (http://www.keiyu.com/)
14. Professional Web Design (Note: In Japanese)
    (http://www.mdn.co.jp/)
15. Research-Based Web Site Design and Usability Guidelines From the
    National Cancer Institute
    (http://usability.gov/guidelines/)
16. The Navigation and Usability Guide by Web Review 1998
    (http://www.webreview.com/1998/05_15/designers/05_15_98_2.shtml)
17. UDIT (Note: In Japanese)
    (http://www.udit-jp.com/)
18. Web Content Accessibility Guidelines 1.0
    (http://www.w3.org/TR/WAI-WEBCONTENT/)
19. Web Style Guide
    (http://books.usableweb.com/display.isbn/0300076754.html)
20. Web Universal Design Guideline
    (http://www.hitachi.co.jp/Div/dc/studio/webud_guideline/)
21. WebTV Design Guide
    (http://developer.webtv.net/design/designguide/Default.htm)

22. Yale C/AIM Web Style Guide
     (http://info.med.yale.edu/caim/manual/contents.html)

REFERENCES

Badre, A., & Badre, A. N. (2002). *Shaping web usability: Interaction design in context.* New York: Addison-Wesley.

Bebit. (2001). *Web usability rulebook* (In Japanese). Tokyo: Bebit.

Blyth, C. S. (1998). *Untangling the web—St. Martin's guide to language and culture on the internet.* New York: St. Martin's.

Brinck, T., Gergle, D., & Wood, S. D. (2001). *Usability for the web: Designing web sites that work.* San Francisco: Kaufmann.

Campbell, N. (2001). *Usability assessment of library-related web sites: Methods and case studies.* Chicago: American Library Association.

Carter, J. (1997, July). The island model of intercultural communication. *SIETAR Japan Newsletter,* Vol. 15.

Cis-z. (2001). *Web interface design guide* (In Japanese). Tokyo: MDN.

Dustin, E., Rashka, J., & McDiarmid, D. (2001). *Quality web systems: performance, security, and usability.* New York: Addison-Wesley.

Deregowski, J. B. (1968). Difficulties in pictorial depth perception in Africa. *British Journal of Psychology, 59,* 195–204.

Esselink, B. (2000). *A practical guide to localization—Language international world directory.* Philadelphia: Benjamins.

Fernandes, T. (1995). *Global interface design.* San Diego, CA: Academic.

Fleming, J. (1998). *Web navigation designing the user experience.* Cambridge, MA: O'Reilly.

Forsythe, C., Grose, E., & Ratner, J. (Eds.). (1998). *Human factors and web development* (1st ed.). Mahwah, NJ: Lawrence Erlbaum Associates.

Hofstede, G. (1991). *Cultures and organizations—Software of the mind.* New York: McGraw-Hill.

Hoshino, A. (1989). Intercultural education and communication (In Japanese). *Intercultural Education, 3,* 4–16.

Hudson, W. (1960). Pictorial depth perception in sub-cultural groups in Africa. *Journal of Social Pychology, 52,* 183–208.

International Organization for Standardization. (1999). *ISO13407—Human-centred design processes for interactive systems.* ISO.

Kilian, C. (1999). *Writing for the web.* North Vancouver, Canada: Self–Counsel.

Krug, S. (2000). *Don't make me think! A common sense approach to web usability.* Indianapolis, IN: New Riders.

Krug, S. (2001). *A common sense approach to web usability* (In Japanese, E. Nakano, Trans.). Tokyo: Softbank.

Kurosu, M. (2001). Cross-cultural user-interface design: What, so what, now what? Four facets of cultural diversity. *HCI International, 2,* 510–513.

Kurosu, M., Itoh, M., & Tokitsu, T. (1999). *User-centered engineering* (In Japanese). Tokyo: Kyoritsu.

Kurosu, M., Motoki, K., & Itoh, Y. (2001). Comparative study of the international software—Do they really fit to the targeted user? *HCI International, 2,* 519–523.

Lopuck, L. (2001). *Web design for dummies.* New York: Hungry Minds.

Lynch, P. J., & Horton, S. (1999). *Web style guide.* New Haven, CT: Yale University Press.

Marcus, A., & Gould, E. W. (2000). Crosscurrents: Cultural Dimensions and Global Web User–Interface Design, Interactions, ACM Publisher, Vol. 7, No. 4, pp. 32–46.

Motoki, K. (2001). *A study on the necessity of high-level localization for the international user interface*. Shizuoka (Japan): Unpublished graduation thesis, Shizuoka University, Faculty of Information.

Nielsen, J. (2000a). *Designing web usability*. Indianapolis, IN: New Riders.

Nielsen, J. (2000b). *Web usability—Secrets for designing the web sites that catch clients* (In Japanese, T. Shinohara, Trans.). Tokyo: MDN.

Nielsen, J., & Del Gäldo, E. M. (Eds.). (1996). *International user interfaces*. New York: Wiley.

Nielsen, J., & Tahir, M. (2001). *Homepage usability: 50 web sites deconstructed*. Indianapolis, IN: New Riders.

Pearrow, M. (2000). *Web site usability handbook*. Cambridge, MA: Charles River Media.

Pearrow, M. (2001). *Web site usability handbook* (In Japanese, K. Modeki, Trans.). Tokyo: Ohmsha.

Pearrow, M. (2002). *The wireless web usability handbook*. Cambridge, MA: Charles River Media.

Roper, J., & Lowe, R. (2002). *Web usability and navigation: A beginner's guide*. New York: Osborne/McGraw-Hill.

Sasaki, H. (1997). *Web design handbook* (In Japanese). Tokyo: AI.

Shade, L. (1998). Cultural aspects. In H. Tamura (Ed.), *Human interface* (In Japanese). Tokyo: Ohm.

Spool, J. M., Snyder, C., Scanlon, T., DeAngelo, T., & Schroeder, W. (1997). *Web site usability: A designer's guide*. North Andover (Mass., USA): User Interface Engineering.

Spool, J. M., Snyder, C., Scanlon, T., DeAngelo, T., & Schroeder, W. (1999). *Web site usability: A designer's guide*. San Francisco: Kaufmann.

Spool, J. M., Snyder, C., Scanlon, T., DeAngelo, T., & Schroeder, W. (2000). *Introduction to web site usability—Consideration on the web site usability* (In Japanese, T. Shinohara, Trans.). Tokyo: Toppan.

Suzuki, K. (1997). *Prologue to cross-cultural psychology* (In Japanese). Tokyo: Brain.

Tajima, J. Y. (2000). *Professional web produce* (In Japanese). Tokyo: SCC.

Taylor, D. (1992). *Global software: Developing applications for the international market*. London: Springer-Verlag.

Tylor, E. B. (1958). *The origin of culture*. Harper & Row.

Uchida, H. (2000). *Basic course on web design—To make it easy to understand and effective* (In Japanese). Tokyo: Visual Design Research Institute.

Yamamoto, Y. (2000). *A better design web page re-design book* (In Japanese). Tokyo: Mainichi Communications.

II

Analyze Phase:
Techniques and Processes

4

Cost-Justifying Web Usability

Deborah J. Mayhew
Deborah J. Mayhew & Associates

Randolph G. Bias
Austin Usability, Inc.

INTRODUCTION

A contract development team was building a Website for a client organization. The Website was to include up-to-date drug information and was intended to be used by physicians as a substitute for the standard desk references they currently use to look up such drug information as side effects, interactions, appropriate uses, and data from clinical trials. The business model for the site was an advertising model. Physicians would visit the site regularly because more current information would be available and that information would be more readily findable on the site, relative to published desk references. Pharmaceutical companies would buy advertising for their drug products on the site because the visitors to the site (physicians) represent their target market. Regular and increasing traffic because of repeat visitors, and new visitors joining based on word of mouth among physicians, would drive up the value of advertising, generating a profit for the client.

The development team generated a prototype design that the client would use to pursue venture capital to support the full-blown development and initial launch and maintenance of the site. The client paid for this prototype development.

Once the prototype was ready, a usability engineering staff was brought on-board to design and conduct a usability test. Eight physicians were paid to fly in to the development center and participate in the test as test users. Several basic search tasks were designed for the physicians to perform. They were pointed to the prototype's home page and left on their own to try to successfully find the drug information that was requested in the first task.

Within 45 seconds of starting their first search task, seven out of the eight physicians in effect gave up and announced, unsolicited, that the site was unusable and that if it were a real site, they would abandon it at that point and never return.

Clearly, if the site had launched as it was prior to this test, not only would an optimal return on investment (ROI) not have been realized, but, in fact, the site would have failed all together and a complete loss of the clients' investment would have resulted. If seven/eighths of all visitors never returned, enough traffic would not have been generated to have motivated advertisers to buy advertising. The entire investment would have been lost.

Instead, the test users were asked to continue with the entire test protocol, and the data generated revealed insights into the problem that was a show stopper on the first task, as well as other problems uncovered in other test tasks. The site was redesigned to eliminate the identified problems. Clearly, the usability test, which had an associated cost, was worth the investment in this case.

This true anecdote illustrates something that distinguishes Websites from commercial software products. In a commercial software product, the buyers discover the usability problems only after they have paid for the product. Often they cannot return it once they have opened the shrink wrap and installed the product. Even if it has a money-back guarantee, they are not likely to return it, and, anyway, perhaps there are not many alternative products on the market with noticeably greater usability.

On a Website, on the other hand, it costs the visitor nothing to make an initial visit. On a Website based on an advertising model, such as the one described, the site sponsor makes nothing at all unless there is sufficient ongoing traffic to attract and keep advertisers. On an e-commerce site, the sponsor makes nothing at all unless the visitors actually find and successfully purchase products, and unless use of the Web channel reduces the costs of other sales channels.

A Website is not a product, and the user does not have to buy it to use it. The Website is just a channel, like a TV show, magazine, or catalog, and if users do not find and repeatedly and successfully use the channel, the investor gets no ROI for having developed it. Thus, usability can absolutely make or break the ROI for a Website even more so than for traditional software products.

It's also true that success competing in the marketplace is even more dependent on relative usability on Websites than is the case with traditional software products or sales channels. Someone wishing to buy a book may be inclined to buy from a particular bricks and mortar bookstore that is easy to get to even if it's not the best bookstore around. On the other hand, if a customer cannot easily find the desired book through the barnesandnoble.com Website, they need not even get out of their chair to shop at a competitor's site instead. It's not enough to simply have a Website that supports direct sales; your site must be more usable than the competition's or you will lose business based on the relative usability of the selling channel alone. For example, 60% of a sample of consumers shopping for travel online stated that if they cannot find what they are looking for quickly and easily on one travel site, they will simply leave and try a competitor's site (Forrester Research, 2000c).

In addition, if you are a catalog order company such as L. L. Bean or Land's End and your product is good but your Website is bad, customers will not use your site and will resort to traditional sales channels (fax and phone) instead. This will result in a poor ROI for a Website that was intended to be justified by relatively low costs compared with those more traditional channels.

USABILITY AS CUSTOMER SERVICE

Site usability is the equivalent of good—or great—customer service. Consider a company's current level of investment in traditional customer service channels. They need to make an equivalent investment in the usability of a site meant to replace or augment traditional channels of customer service. Site usability—like good traditional customer service—ensures that customers can find what they want to buy—an obvious prerequisite to sales. Just as salespeople in stores help you find the book you want, so a good user interface on a bookseller's Website must make browsing and searching easy and successful. Usability—like good customer service—also reduces errors in business transactions and the corresponding costs to fix those errors. For example, just as a good catalog offers accurate pictures of clothing and size charts to minimize the costs of processing returns, a clothing Website must also ensure that customers don't order clothing in the wrong size or color to minimize returns. This might entail providing accurate color images and sizing charts. In addition, usability—like good customer service—motivates customers to choose to use a Website over traditional methods of doing business, ensuring an ROI in the Website itself. Usability helps ensure that customers will return repeatedly to a Website, just as good salespeople help ensure that customers will return to bricks and mortar stores—again contributing to ROI. Also, usability done right in initial development is always cheaper than fixing usability problems identified after launch—or going out of business.

By contrast, lack of usability on an e-commerce site is the equivalent of poor customer service. Imagine having the following conversation with a human telephone customer service representative (CSR):

CSR: Hello. Thank you for calling XYZ Shopping, how can I help you?
Shopper: Hello, I would like to place an order.
CSR: Do you know the name and extension number of the order taker?
Shopper: Of course not! I just want to order something! Can't you take my order?
CSR: Heck no! You need to know who to ask for!

Much later, after finally finding out how to reach the order taker, and then being put on hold repeatedly, and then finally completing giving order information:

Shopper: Oh, I just realized I need to make a change—can you do that?
CSR: If you don't know the name and extension of the order changer, you will
  just have to wait until your order arrives, then send it back and reorder.

Would *you* shop by phone with this company again after such an experience? Unfortunately, this hypothetical interaction is very analogous to many online shopping experiences. E-commerce Websites too often make it very difficult to figure out how to place an order and make changes in midstream, and there are often long periods of being "on hold," waiting for graphics-heavy pages to download.

Thus, there are real risks associated with *not* investing in Web usability. When customers are unsatisfied with the quality of customer service on Websites, this erodes customer loyalty. Even companies with brand loyalty well established through traditional customer service channels may find that loyalty begin to erode when they launch unusable Websites meant to replace traditional customer service channels. Customer dissatisfaction may result from an unacceptable learning curve to accomplish desired goals (as in the example of the physician's drug reference Website cited earlier), from unacceptable task times to accomplish desired goals (e.g., too many clicks or download times that are too long), or from an unacceptable rate of errors and confusion during task completion (e.g., shopping carts abandoned due to lack of shipping costs and time, or failure to notice a final "submit" button). Potential sales may be lost because customers can't find what they want to buy or get the information they need to make buy decisions. Customers may make errors in business transactions that cost time and money to rectify and create customer dissatisfaction. For example, a vendor recently agreed to pay my (Mayhew) return shipping costs because its Website erroneously misled me into ordering an item twice. I will also think twice about using this Website again because I don't want to repeat the error. Customers may refuse to use a Website and return to traditional methods of doing business, reducing the ROI on the Website investment. Customers may defect to competitor companies whose Websites are more usable. In addition, costly rework of a site to fix problems discovered after initial launch may also eat into the ROI.

A Forrester Research, Inc., report, "The Age of Net Pragmatism," (1998) cites the following projected costs of poor Website usability:

• Ill will results from 10 times the monthly traffic, as visitors each communicate their poor experience to 10 friends.
• For every 1 million visitors, 40% don't return due to incomplete content; lost cost of their lifetime value is $2.8 million.
• Customers can't find products; sales underperform by 50% or more.
• $1.5 to $2.1 million is wasted on site redesigns that don't fix the right problem.

The same report quotes University of Maryland's Ben Shneiderman, user experience guru:

- Performance decreases 5% to 10% when designers change the color and position of interface elements. It slows down 25% when developers switch terminology on buttons like "submit" and "search."
- The solution? "Consistency has to be a basic component of the design strategy." Clumsy navigation is one reason why, on average, only 2.7% of site lookers turn into buyers.

Sometimes lack of usability not only erodes ROI but actually eliminates it altogether. While not a Website example, the following example is nevertheless compelling. Nielsen (Nielsen 1993, p. 84) reports a case involving the interface to the installation process for an upgrade of a spreadsheet package. When the upgrade was shipped, customers needed an average of two 20-minute calls each to customer support to correctly install the upgrade. Support calls to the vendor cost them an average of $20 per 5 minutes to service; thus, the support cost per customer for this product was about $160. Unfortunately, the profit margin on the upgrade product was only $70 per customer; thus, not only was the ROI on the product eroded, the upgrade product actually *cost* the vendor nearly $100 per customer— all due to usability problems that probably could have been detected and fixed fairly cheaply prior to releasing the upgrade product.

While this example involved a traditional software product rather than a Website, one can easily imagine analogous situations on Websites, when customers who cannot figure out how to use an e-commerce Website must call traditional customer service to complete their purchases. In fact, a whole new type of Internet-based software product has emerged to handle this situation: Live customer service buttons ("Call Back" buttons) have begun appearing on Websites so that when a visitor gets lost or confused or cannot find something, they can get connected immediately through phone or text chat channels to live customer service reps who can share a screen with them and guide them through the site. There is a certain irony here—the implication is that the site itself requires customer support when in fact it was meant to *be* customer support! It might be cheaper and more sensible to invest in improving Website usability than to provide human customer support through Call Back buttons to help visitors use the site successfully. In fact, some e-commerce Websites have failed completely due to poor usability and have been shut down (Forrester Research, 2000a, cites boo.com and levi.com as examples).

The Graphic, Visualization & Usability Center (GVU) of the Georgia Institute of Technology in Atlanta, Georgia, publishes periodic surveys of WWW users. In GVU's tenth WWW user survey (May 14, 1999, http://www.gvu.gatech.edu/user_surveys), 5,000 users were surveyed. When asked to check any of 19 listed problems using the Web as the biggest, a large number of respondents checked problems related to site navigation—an important aspect of site usability:

- 45.4%: "Not being able to find the information I am looking for."
- 30.0%: "Not being able to find a page I know is out there."
- 16.6%: "Not being able to return to a page I once visited."

Survey users were also asked: "Which of the following features are *most important* to you personally, when shopping, or considering shopping, on the web. (Check all that apply.)" There were 14 possible responses, and 4 were directly related to ease of use. These were (with the percentage of respondents who checked each one):

- Ease of placing orders (76.6%)
- Ease of canceling orders (47.0%)
- Easy payment procedures (54.4%)
- Ease of contacting the vendor (53.8%)

Clearly, when shopping on the Web, ease of use is very important to most users.

When asked if they ever had dissatisfying Web shopping experiences leading to leaving one site for another, and if so, what the specific sources of dissatisfaction were, survey users responded as follows:

- 83.8% said yes. In particular:
  66.8% said, "The site was disorganized/confusing."
  59.0% said, "I could not find what I was looking for."
  37.9% said, "The individual pages were poorly designed."

These are all usability issues, leading users to abandon sites for competitive sites when shopping.

Twenty-eight percent of survey users stated that on average it took more than 10 minutes to find something they were looking for. Of all respondents, 23.4% said that if they could not find something within 10 minutes they would give up. Ten minutes is a *very* long time! The time to find a desired item could be far less with better usability. Clearly, sites are losing some users due to the amount of time it takes to find things. Time to find a desired item is—and will become even more so—an aspect of competitive edge across competing Websites.

Forrester reports a buy-to-look ratio of only about 2% on e-commerce Websites (Forrester Research, 2000a), and although part of this may be due to other factors, usability is surely a large factor in this poor performance of Websites as a sales channel.

Another Forrester Report (Forrester Research, 2001) cites other compelling statistics from a variety of studies. In one study, 65% of shopping attempts at a set of prominent e-commerce sites ended in failure. In another, 42% of users were unable to complete the job application process on a set of corporate sites. Another study showed that journalists could find answers to only 60% of their questions on corporate public relations sites. Still another study showed that adding certain product information to e-commerce sites reduced product-related inquiries by more than 20%. Finally, Forrester points out that when Amazon.com let customer service slip in 2000, they fell behind their competitors in three categories for the first time.

In this same report, Forrester notes that the companies it interviewed spent between $100,000 and $1 million on site redesigns, but few had any sense of which specific design changes paid off, if any. Clearly, usability engineering programs that are specifically aimed at such goals as increased success rate of shopping attempts, job application completion, and information requests, could have had a profound impact on the ROI of these redesign projects.

As illustrated by all these examples and statistics, usability engineering could clearly have a profound impact on ROI in the case of Websites, and even more dramatic cost justification cases can be made for usability engineering programs during Web development as compared to during traditional software development projects. A profit is made whenever a commercial software product is sold, regardless of how productive users are once they start to use that software product, but there simply is no ROI until and unless visitors to Websites complete their tasks efficiently and successfully, and then return again and again.

A FRAMEWORK

This chapter presents a framework for cost-justifying usability engineering efforts for Web development projects by describing how to go about calculating the costs and estimating the benefits of the Usability Engineering Lifecycle tasks that can potentially be applied (Mayhew, 1999; also see Chapter 1 of this book for an overview). The subject of cost-justifying usability engineering in general is covered extensively in a book called *Cost-Justifying Usability* (Bias & Mayhew, 1994). In particular, Chapter 2 in that book elaborates on the simple framework offered in this chapter. Other chapters in that book offer further discussion on the topic and case studies. Readers interested in learning more about how, when, and why to cost-justify usability engineering efforts are referred to that book for more detail than can be provided here. The purpose of this chapter is to provide an overview of how that more general cost justification technique can be applied to Web development projects in particular.

Purposes for Cost Justification

There are two possible purposes for conducting a cost–benefit analysis of usability engineering efforts for Web development projects: (1) to "sell" usability engineering in general, and (2) to plan for a usability engineering program for a particular development project.

Very general cost–benefit analyses of hypothetical usability engineering efforts can be prepared as a strategy to win general support for trying out usability engineering techniques in a Web development organization. When an organization has no experience with usability engineering, cost–benefit analyses can help to win resources to experiment with it for the first time.

In more mature organizations, cost–benefit analyses can be used to plan an optimal usability engineering effort for a particular Web development project.

To help settle on a final usability engineering plan for a specific development project, you could start by calculating the costs of the most aggressive usability engineering program that you might like to implement, including rigorous techniques for all lifecycle tasks (see Chapter 1). Then you could estimate benefits, using very conservative estimates of benefits. If benefits still outweigh costs dramatically, as they usually do when critical parameters are favorable, then you could easily make a good argument for even the most aggressive usability engineering program, because only the most conservative claims concerning potential benefits have been made and as such can be defended easily. In fact, you can then go back and redo the benefits estimates using more aggressive yet still realistic benefit assumptions and show that in all likelihood an even more dramatic overall benefit will be realized, even from a significant investment in usability engineering.

If, however, benefits and costs in the initial analysis seem to match up fairly closely, then you will want to scale back your initial, aggressive usability engineering program, maybe even to just a bare-bones plan. For example, perhaps you should plan to do only a "quick and dirty" user profile by interviewing user management instead of conducting a full blown user survey, a very quick and dirty task analysis consisting of just a few rounds of contextual observations and interviews with users rather than many and then just one iterative cycle of usability testing on a complete detailed design rather than iterations at several stages of design to catch major flaws and be sure that you have achieved the predicted benefits. You are still likely to realize the very conservative assumptions made concerning benefits with just these minimal usability techniques, and thus you could predict with confidence a healthy ROI from a more minimal approach to usability engineering. This is probably wiser in the long run than making overly optimistic claims concerning potential benefits, spending a lot to achieve them, and then perhaps not achieving them and losing credibility. Thus, you can use a cost–benefit analysis in this fashion, to develop a sensible usability engineering effort for a Web development project, likely to pay off as predicted.

When an organization is first experimenting with usability engineering techniques and is still skeptical about their value, it is wise to make very conservative cost–benefit arguments, based on a relatively low cost usability engineering effort and very modest estimates of benefits, and then to show after the fact that much larger benefits were in fact realized. Once an organization has several positive experiences investing in usability engineering, it will be more receptive to more aggressive proposals for usability engineering programs and to more optimistic benefits estimates in the cost–benefit analyses that argue for them. See Bias and Mayhew (1994, Chapter 2) for additional concrete examples of how to use cost–benefit analyses to tailor usability engineering programs.

General Approach

To cost-justify a usability engineering plan, you simply adapt a very generic and widely used cost–benefit analysis technique. Having laid out a detailed usability project plan based on lifecycle tasks (see Mayhew, 1999, and Chapter 1 of this book), it is a fairly straightforward matter to calculate the costs of the plan. Then you need to calculate the benefits. This is a little trickier and where the adaptation of the generic analysis comes into play. Then, you simply compare costs to benefits to find out if and to what extent the benefits outweigh the costs. If they do to a satisfactory extent, then you have cost-justified the planned effort.

More specifically, first a usability engineering plan is laid out. The plan specifies particular techniques to employ for each lifecycle task, breaks the techniques down into steps, and specifies the personnel hours and equipment costs for each step. The cost of each task is then calculated by multiplying the total number of hours for each type of personnel by their effective hourly wage (fully loaded, as we explain below) and adding in any equipment and other costs. Then the costs from all tasks are summed to arrive at a total cost for the plan.

Next, overall benefits are predicted by selecting relevant benefit categories, calculating expected benefits by plugging project-specific parameters and assumptions in to benefit formulas, and summing benefits across categories. The list of possible benefits to consider is long, as usability engineering leads to tangible benefits to all concerned. The site development team realizes savings, as problems are identified early, when they are cheap to fix. The customer support team realizes a reduced call support burden. More usable sites will have higher buy-to-look ratios, a lower rate of abandoned shopping carts due to errors and confusion, and increased return visits. More usable sites will also have fewer failed searches, fewer errors, and thus higher user productivity. Also, for intranet sites or Web-based applications, more usable sites will have lower associated user training costs.

The potential benefit categories selected in a particular cost–benefit analysis will depend on the basic business model for the site. Sample benefit categories potentially relevant to e-commerce sites include

- Increased buy-to-look ratios.
- Decreased abandoned shopping carts.
- Increased return visits.
- Decreased costs of other sales channels.
- Decreased use of the Call Back button.
- Decreased development costs.

Sample benefit categories potentially relevant to sites based on an advertising business model include

- Increased number of visitors.
- Increased number of return visitors.

- Increased length of visits.
- Increased click through on ads.
- Decreased failed searches.
- Decreased development costs.

Sample benefit categories potentially relevant to product information sites include

- Increased sales leads.
- Decreased development costs.

Sample benefit categories potentially relevant to customer service sites include

- Decreased costs of traditional customer service channels.
- Decreased development costs.

Sample benefit categories potentially relevant to intranet sites supporting internal business users include

- Decreased user training costs.
- Increased user productivity.
- Decreased user errors.
- Decreased user support.
- Decreased development costs.

Note that the relevant benefit categories for different types of sites vary somewhat. Thus, in a cost–benefit analysis, one wants to focus attention on the potential benefits that are of most relevance to the bottom-line business goals for the site.

Also note that these benefits represent just a sample of those that might be relevant in these types of sites and does not address other possible benefits of usability in other types of sites. Others might be included as appropriate, given the business goals of the site sponsor and the primary concerns of the audience and could be calculated in a similar fashion as these are (see "Sample Cost–Benefit Analyses," which follows). Finally, overall benefits are compared to overall costs to see if, and to what extent, the overall usability engineering plan is justified.

When usability practitioners are invited to participate in projects already in progress, which is often the case for external consultants, they have less chance of including all Usability Engineering Lifecycle tasks and of influencing overall schedules and budgets. They are more likely to have to live within already committed to schedules, platforms, and system architectures, use shortcut techniques for Usability Engineering Lifecycle tasks, and minimally impact budgets. Nevertheless, it is almost always possible to create a usability engineering plan that will make a significant contribution to a project even when one comes in relatively late. In addition, you can use the cost–benefit analysis technique to prepare and support usability engineering plans that involve only parts of the overall Usability Engineering Lifecycle and only shortcut techniques for tasks within it.

Sample Cost–Benefit Analyses

Let us consider a hypothetical usability engineering plan in the context of two different Web development projects and then see how you would conduct cost–benefit analyses of that plan for each project.

An E-commerce Site

Imagine that a Web development organization is planning to redesign an existing e-commerce site that is not producing the ROI hoped for. Traffic statistics are available from the current site.

First, the final results of such an analysis are presented. Then, in the steps below, the derivation of the final results are shown.

Table 4.1 shows the overall calculation of the cost of a usability engineering plan proposed by the project usability engineer. The first column identifies the overall project phase. The second column identifies which Usability Engineering Lifecycle tasks (see Chapter 1) are planned in each phase. The third, fourth, and fifth columns identify the number of hours of usability engineers, developers, and users, respectively, that are required to complete each task. The last column then summarizes the total cost of each task. A grand cost total for the whole plan is given at the bottom of the table.

The project usability engineer also estimated that the usability engineering plan in Table 4.1 will produce a new site design that will produce the benefits given in Table 4.2 every month.

TABLE 4.1
Cost of Usability Engineering Plan

Phase	Task	Usability Engineer Hours @ $175	Developer Hours @ $175	User Hours @ $25	Total Cost
Requirements analysis	User profile	60		25	$11,125
	Task analysis	80		40	$15,000
	Platform constraints	8	8		$2,800
	Usability goal setting	20			$3,500
Design/test/ develop	Information architecture	80			$14,000
	Conceptual model design	80			$14,000
	Paper prototype development	20			$3,500
	Usability test	160		16	$28,400
	Screen design standards	80			$14,000
	Live prototype development	20	80		$17,500
	Usability test	160		16	$28,400
	Detailed user interface design	80			$14,000
	Usability test	160		16	$28,400
	Totals	1008	88	113	$194,625

TABLE 4.2
Expected Monthly Benefits for an E-business Site

Benefit Category	Benefit Value Per Month
Decreased abandoned shopping carts	$31,250.00
Increased buy-to-look ratio	$31,250.00
Decreased use of Call Back button	$4,166.80
Total	$66,666.80

Comparing these benefits and costs, the project usability engineer argued that the proposed usability engineering plan will pay for itself in the first 3 months after launch:

Benefits per month = $66,666.80
One-time cost = $194,625.00

Payoff period = 2.92 months

As the new site is expected to have a lifetime of something much longer than 3 months, the project usability engineer expected her or his plan to be approved based on this cost justification.

Note that the analyses offered here do not consider the time value of money—that is, the money for the costs is spent at some point in time, whereas the benefits come later in time. Also, if the money was *not* spent on the costs but was invested instead, this money would likely increase in value. Usually, the benefits of usability are so robust that these more sophisticated financial considerations aren't necessary. However, if needed, calculations based on the time value of money are presented in Bias, Mayhew, and Upmanyu (2002) and also in Bias and Mayhew (1994). What follows is a step-by-step description of how the project usability engineer arrived at the above final results.

1. Start with the Usability Engineering Plan.   If this has not already been done, it is the first step in conducting a cost–benefit analysis. The usability engineering plan identifies which Usability Engineering Lifecycle tasks and techniques will be employed (see Chapter 1) and breaks them down into required staff and hours. Costs can then be computed for these tasks in the next two steps.

2. Establish Analysis Parameters.   Many of the calculations for both planned costs and estimated benefits are based on project-specific parameters. These should be established and documented before proceeding with the analysis. Sample parameters are given below for our hypothetical project:

- The site is an e-business site (vs. a product information site or a site based on an advertising model).

- The current site gets an average of 125,000 visitors per month (vs., for example, 500,000 or 10,000 users).
- The current buy-to-look ratio is 2%.
- The current average profit margin on each online purchase is $25.
- The current rate of use of the Call Back button is 2%.
- Average length of each servicing of use of the Call Back button is 5 minutes.
- Users are paid to participate in usability engineering tasks at an hourly rate of $25.
- The customer support fully loaded hourly wage is $40.
- The usability engineering and developer fully loaded hourly wage is $175.
- A usability lab or sufficient resources are in place and already paid for.

It should be emphasized that when using the general cost–benefit analysis technique illustrated here, these particular parameter values should not be assumed. The particular parameter values of your project and organization should be substituted for those above. They will almost certainly be different from the parameters used in this example.

You should note, in general, that certain parameters in a cost–benefit analysis have a major impact on the magnitude of potential benefits. For example, when considering productivity—of primary interest in the case of intranet sites for internal business users—the critical parameters are the number of users and the volume of transactions and to some extent also the user's fully loaded hourly wage. When there is a large number of users and/or a high volume of transactions, even very small performance advantages (and low hourly wages) in an optimized interface quickly add up to significant overall benefits. On the other hand, when there is a small number of potential users and/or a low volume of transactions, benefits may not add up to much even when the potential per transaction performance advantage seems significant, and the user's hourly wage is higher.

For example, consider the following two scenarios. First, imagine an intranet site that supports internal business users. Let's say there are 5,000 users and 120 transactions per day per user. In this case, even a half second performance improvement per transaction adds up:

$$5{,}000 \text{ users} \times 120 \text{ transactions} \times 230 \text{ days per year} \times 1/2 \text{ second} = 19{,}167 \text{ hours}$$

If the user's hourly rate is $15, the savings are

$$19{,}167 \text{ hours} \times \$15 = \$287{,}505$$

This is a pretty dramatic benefit for a tiny improvement on a per transaction basis. On the other hand, if there were only 25 users, and they were infrequent users with only 12 transactions per day, even if a per transaction benefit of 1 minute could be realized, the overall benefit would only be:

$$25 \text{ users} \times 12 \text{ transactions} \times 230 \text{ days per year} \times 1 \text{ minute} = 1{,}150 \text{ hours}$$

At $15 per hour, the overall productivity benefit is only:

$$1{,}150 \text{ hours} \times \$15 = \$17{,}250$$

Thus, in the case of productivity benefits, costs associated with optimizing the user interface are more likely to pay off when there are more users and more transactions. In the case of an e-commerce site, the critical parameters are usually volume of visitors and profit margin per purchase. If the profit margin per online purchase is low, then a very large number of additional purchases would have to be achieved due to usability alone for the usability engineering costs to pay off. On the other hand, if the profit margin per online purchase is high, then only a small number of increased purchases would be necessary to pay for the usability engineering program because of improved usability. Thus, these critical parameters are going to directly determine how much can be invested in usability and still pay off. See Bias and Mayhew (1994, Chapter 2) for more discussion of analysis parameters.

3. Calculate the Cost of Each Usability Engineering Lifecycle Task in the Usability Engineering Plan. Given a usability engineering plan, the number of hours required by each type of necessary staff can be estimated for each task and technique in the plan. Once you can estimate the number of hours required for each usability engineering task and technique, you then simply multiply the total number of hours of each type of staff required for each task by the fully loaded hourly wage of that type of staff and sum across the different staff types. Additional costs, such as equipment and supplies, should also be estimated and added in for each task. This is how the individual total task costs in Table 4.1 were calculated.

The numbers of hours estimated for each task and technique in the cost summary table were not pulled out of a hat—they were based on many years of experience. For example, the 160 hours estimated to conduct a usability test was derived as follows:

1.	Design and develop test materials	32 hours
2.	Design and assemble the test environment	8
3.	Run a pilot test	8
4.	Revise test tasks and materials	8
5.	Run the test and collect data	32
6.	Summarize and interpret the data, draw conclusions	32
7.	Document and present results	40
	Total	160

Similar sample breakdowns of estimates for all Usability Engineering Lifecycle tasks and techniques can be found in Mayhew (1999).

Fully loaded hourly wages are calculated by adding together the costs of salary, benefits, office space, equipment, and other facilities for a type of personnel and dividing this by the number of hours worked each year by that personnel type. If outside consultants or contractors are used, their simple hourly rate plus travel expenses would apply.

The hourly rate for users used in this cost estimate was $25. This was not based on a typical user's fully loaded hourly wage at their job, which it would be in the case of a cost justification of traditional software development for internal users or intranet development for internal users. Instead, it was premised on the assumption that in the case of an e-commerce Website, test users will have to be recruited from the general public to participate in usability engineering tasks and techniques, and that they would be paid at a rate of $25 an hour for their time.

The hourly rate for usability engineering staff was based on an average of the current salaries of senior level internal usability engineering staff and external consultants. The hourly rate of developers was similarly estimated.

Remember that both the hourly wage figures and the predicted hours per task used to generate the sample analyses here are just examples. You would have to use the actual fully loaded hourly rates of personnel in your own organization (or the rates of consultants being hired) and the expected time of your own usability engineers to complete each task based on their experience and the techniques they plan to use to carry out your own analysis.

4. Select Relevant Benefit Categories.   Because this is an e-business site, only certain benefit categories are of relevance to the business goals of the site. For another kind of site, different benefit categories would be selected.

In this hypothetical case, the project usability engineer decided to include the following benefits:

- Increased buy-to-look ratio.
- Decreased abandoned shopping carts.
- Decreased use of the Call Back button.

These were selected because she or he knew that they would be of most relevance to the audience for the analysis, the business sponsors of the site. There may be other very real potential benefits of this usability engineering plan, but these were chosen only for simplicity and to make a conservative estimate of benefits.

As compared to the existing site design, the usability engineer anticipated that in the course of redesign the usability engineering effort would decrease the amount of abandoned shopping carts by ensuring that the checkout process is clear, efficient, provides all the right information at the right time, and does not bother users with tedious data entry of information they do not want or need to provide. She or he expected to improve the buy-to-look ratio by ensuring that the right product information is contained in the site and that navigation to find products is efficient and always successful. Use of the Call Back button was expected to decrease by making the information architecture match users expectations and by designing and validating a clear conceptual model so that navigation of and interactions with the site are intuitively obvious. Accomplishing all these things depends on conducting the requirements analysis and testing activities in the proposed plan, as well as on applying general user interface design expertise.

5. *Estimate Benefits.*   Next, the project usability engineer estimated the magnitude of each benefit that would be realized—compared to the current site, which is being redesigned—*if* the usability engineering plan (with its associated costs) were implemented. Thus, for example, she or he estimated how much higher the buy-to-look ratio on the site would be if it were reengineered for usability as compared to the existing site.

To estimate each benefit, choose a unit of measurement for the benefit, such as the average purchase profit margin in the case of the increased buy-to-look ratio benefit or the average cost of customer support time spent servicing each use of the Call Back button. Then—and this is the tricky part—make an assumption concerning the magnitude of the benefit for each unit of measurement, for example, a 1% increase in buy-to-look ratio or a 1% decrease in the use rate of the Call Back button. (Tips on how to make these key assumptions are discussed later.) Finally, calculate the total benefit in each category based on the unit of measurement, key parameters, and your assumptions about magnitudes of benefit. When the unit of measurement is time, benefits can be expressed first in units of time and then converted to dollars, given the value of time.

Remember that our hypothetical project involves development of an e-commerce site. Based in part on her or his in-house experience, our project usability engineer made the following assumptions:

- Buy-to-look ratio will increase by 1% of total visitors.
- Abandoned shopping carts will decrease by 1% of total visitors.
- Use of the Call Back button will decrease by 1% of total visitors.

Then, benefits in each of these categories were calculated as follows:

Increased Buy-to-Look Ratio:
　　1% more visitors will decide to buy and will successfully make a purchase.
　　125,000 visitors per month × 1% = 1,250 more purchases.
　　1,250 purchases at a profit margin of $25 = $31,250 per month.

Decreased Abandoned Shopping Carts:
　　1% more visitors who would have decided to buy anyway will now complete
　　　checkout successfully and make a purchase.
　　125,000 visitors per month × 1% = 1,250 more purchases.
　　1,250 purchases at a profit margin of $25 = $31,250 per month.

Decreased Use of "Call Back" Button:
1% fewer visitors will need to use the Call Back button.
125,000 visitors per month × 1% = 1,250 fewer calls.
1,250 calls at 5 minutes each = 104.17 hours.
104.17 hours at $40 = $4,166.80 per month.

The usability engineer based her or his assumptions on statistics available in the literature, such as those presented in the introduction to this chapter. In particular, she or he began with the often quoted average e-commerce Website buy-to-look ratio of 2% to 3%. She or he then based an assumption that this rate could be improved by a minimum of 2% (1% from improving the product search process and 1% from improving the checkout process) through usability engineering techniques, on statistics offered as average by a Forrester report called "Get ROI from Design (2001). This report suggested that it would be typical for as much as 5% of online shoppers to fail to find the product and offer they are looking for (other statistics suggest as much as 45% may experience this problem) and for over 50% of shoppers that do find a product they would like to buy to bail out during checkout due to confusion over forced registration, being asked to give a credit card number before being shown total costs, and because they are not convinced that this is the best available offer. She or he based an assumption of reduced use of the Call Back button by 1% on statistics cited earlier suggesting that as much as 20% of site users typically call in to get more information. Most of us have experienced these problems and would have little argument with the idea that they are typical. Given these statistics, the assumptions made above seem very modest.

The basic assumption of a cost–benefit analysis of a usability engineering plan is that the improved user interfaces that are achieved through usability engineering techniques will result in such tangible, measurable benefits, like those calculated in this hypothetical example.

The audience for the analysis is asked to accept these assumptions of certain estimated, quantified benefits as reasonable and likely minimum benefits rather than as precise, proven, guaranteed benefits. Proof simply does not exist for each specific Website that an optimal user interface will provide some specific, reliable advantage over some other user interface that would—or has—resulted in the absence of a usability engineering plan.

How can you generate—and convince your audience to accept—the inherent assumptions in the benefits you claim in a given cost–benefit analysis? First, it should be pointed out that a cost–benefit analysis for any purpose must make certain assumptions that are really only predictions of the likely outcome of investments of various sorts. The whole point of a cost–benefit analysis is to try to evaluate in advance, in a situation in which there is some element of uncertainty, the likelihood that an investment will pay off. The trick is in basing the predictions of uncertainties on a firm foundation of known facts. In the case of a cost–benefit analysis of a planned usability engineering program, there are several foundations on which to build realistic predictions of benefits.

First, there is ample published research that shows measurable and significant performance advantages of specific user interface design alternatives (as compared to other alternatives) under certain circumstances.

For example, in the quote cited earlier from a Forrester Research, Inc., report (1998), Ben Shneiderman states that "Performance decreases 5% to 10% when designers change the color and position of interface elements. It slows down 25% when developers switch terminology on buttons like 'submit' and 'search.'"

In another example, Lee and Bowers (1997) studied the conditions under which the most learning occurred from presented information and found that compared to a control group that had no learning opportunity:

- Hearing spoken text and looking at graphics resulted in 91% more learning.
- Looking at graphics alone—63% more.
- Reading printed text plus looking at graphics—56% more.
- Listening to spoken text, reading text, and looking at graphics—46% more.
- Hearing spoken text plus reading printed text—32% more.
- Reading printed text alone—12% more.
- Hearing spoken text alone—7% more.

Clearly, how information is presented makes a significant difference in what is retained.

The research does not provide simple generic answers to design questions. However, what the research does provide are general ideas of the magnitude of performance differences that can occur between optimal and suboptimal interface design alternatives. The basic benefit assumptions made in any cost–benefit analysis can thus be generated and defended by referring to the wide body of published research data that exists. From these studies, you can extrapolate to make some reasonable predictions about the order of magnitude of differences you might expect to see in Website user interfaces optimized through usability engineering techniques (see Bias & Mayhew, 1994, Chapter 2, for a review and analysis of some of the literature for the purpose of defending benefit assumptions).

Besides citing relevant research literature, there are other ways to arrive at and defend your benefit assumptions. Actual case histories of the benefits achieved as a result of applying usability engineering techniques are very useful in helping to defend the benefits assumptions of a particular cost–benefit analysis. A few published case histories exist (e.g., Bias & Mayhew, 1994; Karat, 1989). Wixon and Wilson (1997) and Whiteside, Bennet, and Holtzblatt (1988) report that across their experience with many projects over many years, they find that they average an overall performance improvement of about 30% when at least 70% to 80% of the problems they identify during usability testing are addressed by designers.

But even anecdotes are useful. For example, a colleague working at a vendor company once told me (Mayhew) that they had compared customer support calls on a product for which they had recently developed and introduced a new, usability-engineered release. Calls to customer support after the new release decreased by 30%. This savings greatly outweighed the cost of the usability engineering effort. For another, even more dramatic anecdote, see again the example cited earlier by Nielsen (1993, p. 84), regarding the upgrade to a spreadsheet application.

Finally, experienced usability engineers can draw on their own general experience evaluating and testing software user interfaces and their specific experience with a particular development organization. Familiarity with typical interface designs from a particular development organization allows the usability engineer to decide how much improvement to expect from applying usability engineering techniques in that organization. If the designers are generally untrained and inexperienced in interface design and typically design poor interfaces, the usability engineer would feel comfortable and justified defending more aggressive benefits claims. On the other hand, if the usability engineer knows the development organization to be quite experienced and effective in interface design, then more conservative estimates of benefits would be appropriate, on the assumption that usability engineering techniques will result in fine-tuning of the interface but not radical improvements. The usability engineer can also assess typical interfaces from a given development organization against well-known and accepted design principles, usability test results, and research literature to help defend the assumptions made when estimating benefits.

In general, it is usually wise to make very conservative benefit assumptions. This is because any cost–benefit analysis has an intended audience, which must be convinced that benefits outweigh costs. Assumptions that are very conservative are less likely to be challenged by the relevant audience, thus increasing the likelihood of acceptance of the analysis conclusions. In addition, conservative benefits assumptions help to manage expectations. It is always better to achieve a greater benefit than was predicted than to achieve less benefit, even if it still outweighs the costs. Having underestimated benefits will make future cost–benefit analyses more credible and more readily accepted.

When each relevant benefit has been calculated for a common unit of time (e.g., per month or per year), then it is time to add up all benefit category estimates for a benefit total.

6. Compare Costs to Benefits.   Recall that in our hypothetical project, the project usability engineer compared costs to benefits, and this is what was found:

Benefits per month = $66,666.80
One-time cost = $194,625.00
Payoff period = 2.92 months

Our project usability engineer's initial usability engineering plan appears to be well justified. It was a fairly aggressive plan, in that it included all lifecycle tasks and moderate to very rigorous techniques for each task, and the benefit assumptions were fairly conservative. Given the very clear net benefit, she or he would be wise to stick with this aggressive plan and submit it to project management for approval.

If the estimated payoff period had been long or in fact there was no reasonable payoff period, then the project usability engineer would have been well-advised to go back and rethink the plan, scaling back on the rigorousness of techniques for certain tasks and even eliminating some tasks (for example, collapsing the design process from three to two or even one design level) to reduce the costs.

A Product Information Site

This example is based on a hypothetical scenario given in a Forrester report called "Get ROI from Design," (2001). It involves an automobile manufacturing company that has put up a Website that allows customers to get information about the features of the different models of cars it offers and the options available on those cars. It allows users to configure a base model with options of their choice and get sticker price information. Users cannot purchase a car online through this Website—it is meant to generate leads and point users to dealerships and salespeople in their area.

1. Start with the Usability Engineering Plan. In this example, we will again start with the same assumed plan as in the e-commerce site example (see Table 4.1).

2. Establish Analysis Parameters. Sample parameters are given below for this example. Again, we are assuming there is an existing site with known traffic statistics and that the project involves a redesign:

- The site is a product information site (vs. an e-business site or a site based on an advertising model).
- The current site gets an average of 500,000 visitors per month (vs., for example, 125,000 or 10,000 visitors).
- Currently, 1% of visitors result in a concrete lead.
- Currently, 10% of leads generate a sale.
- The profit on a sale averages $300.
- Users are paid to participate in usability engineering tasks at an hourly rate of $25.
- The usability engineering and developer fully loaded hourly wage is $175.
- The usability lab or sufficient resources are in place and already paid for.

3. Calculate the Cost of Each Usability Engineering Lifecycle Task in the Usability Engineering Plan. We can use the same cost calculations as before, shown in Table 4.1.

4. Select Relevant Benefit Categories. Since this is a product information site, only certain benefit categories relevant to the business goals of this redesign project. The project usability engineer decides to include only the following benefit category: increased lead generation.

This category was selected because she or he knows it will be most relevant to the audience for the analysis, the business sponsors of the site. There may be other very real potential benefits of the usability engineering plan, but she or he chose only this one for simplicity and to make a conservative estimate of benefits.

As compared to the existing site design, the usability engineer anticipated that in the course of redesign the usability engineering effort will increase leads by ensuring that visitors can find basic information and successfully configure models with options. Accomplishing this will depend on conducting the requirements analysis and testing activities in the proposed plan, as well as on applying general user interface design expertise.

5. Estimate Benefits. Next, the project usability engineer estimated the magnitude of the benefit that would be realized—relative to the current site, which is being redesigned—*if* the usability engineering plan (with its associated costs) were implemented. Thus, in this case, she or he estimated how much higher the lead generation rate on the site would be if it were reengineered for usability as compared to the existing site.

The estimated benefit was calculated as follows:

Increased Lead-Generation Rate:
1% more visitors will generate a lead.
500,000 visitors per month $\times$ 1% = 5,000 more leads.
10% of these new leads resulting in a sale = 500 more sales.
500 more sales at a profit margin of $300 = $150,000 per month.

6. Compare Costs to Benefits. Next, the usability engineer compared benefits and costs to determine the payoff period:

Benefits per month = $150,000
One-time cost = $194,625

Payoff period = 1.3 months

Again, our project usability engineer's initial usability engineering plan appears to be well justified. It was a fairly aggressive plan, in that it included all lifecycle tasks and moderate to very rigorous techniques for each task, and the benefit assumptions were fairly conservative. Given the very short estimated payoff period, she or he would be wise to stick with this aggressive plan and submit it to project management for approval. In fact, based on the very modest assumption of a 1% increase in lead generation, it might be well advised to redesign the usability engineering plan to be even more thorough and aggressive because increased benefits that might be realized by a more rigorous approach will likely be more than compensated for. In this case, she or he might be well advised to increase the level of effort of the requirements analysis tasks, which usually have a high payoff.

Returning to the Actuals

Time may not allow it—you'll be eager to move on to the next project—but there may be value in gathering the data on the actual benefits that resulted from your usability engineering efforts after the fact. How much did the buy-to-look ratio increase? How many fewer calls to the help desk (per visitor) were there? How much was the return visitor percentage increased? It will be hard to attribute all changes to the improved usability of the site—perhaps a new ad campaign contributed to driving more visitors to the site. But reasonable estimates can be offered to justify the cost–benefit approach and the usability engineering program in general. Also, as an in-between step, if you are able to perform multiple rounds of testing, you'll have actual numbers in the second round that will demonstrate, presumably, actual benefits due to usability engineering.

SUMMARY

All the above examples are based on a simple subset of all actual costs and benefits, and very simple and basic assumptions regarding the value of money over time. More complex and sophisticated analyses can be calculated (see Bias & Mayhew, 1994; Bias et al., 2002, Chapter 64). However, usually a simple and straightforward analysis of the type offered in the examples above is sufficient for the purpose of winning funding for usability engineering investments in general or planning appropriate usability engineering programs for specific development projects.

One key to applying cost justification to planning usability engineering programs for Web development projects is the generation of Web traffic statistics that are relevant and useful. In the Forrester Research report called "Measuring Web Success" (1999), it is pointed out that typical Web traffic statistics don't really tell you what you want to know. For example, knowing how many visitors to a product information site there were this month and how many of them viewed each product does not tell you how many customers were satisfied with the information provided. Similarly, knowing how many visitors to an e-commerce site made a purchase this month and what the average purchase price was does not tell you why more shoppers did not buy. Also, knowing how many visits a customer service site received and how many e-mails were received and responded to by live customer service reps does not tell you how many customer support issues were resolved online. When we have traffic software that can generate more specific and integrated statistics that allow these kinds of inferences, it will become much easier to validate cost justification estimates after the fact and make future cost–benefit analyses much more credible. Ideally, says this report:

> Managers will have self-service access to relevant information: Human resources will get reports on job seekers, web designers will see detail on failed navigations, and marketing will receive customer retention metrics. . . . For example, a marketing director trying to determine why look-to-buy ratios plummet after 6:00 pm might discover that more evening users connect through dial-up modems that produce a slower and more frustrating shopping experience.

This report also points out that cross-channel statistics that track a given customer are just as important—if not more so—than simple Web traffic statistics. For example, if it could be captured that a particular user visited a customer service Website but then shortly thereafter called the customer service hot line, this might point to some flaw or omission in the information in the online customer service channel. In addition, if the customer's "tracks" through the Website as well as the nature of the phone conversation could also be captured and integrated, inferences might be made about exactly how the Website failed to address the customer's need.

This kind of very specific traffic tracking starts to sound very much like a sort of ongoing automated usability test, which could generate data that would point very directly to specific user interface design solutions, making the case for usability engineering even stronger.

One might look at typical Web development time frames from the early years of Web development and notice that while the average Web development project back then might have taken all of 8 to 12 weeks from planning to launch, the usability engineering program laid out in the sample analyses in this chapter would itself take an absolute minimum of 7 weeks. How can it make sense to propose a usability engineering program—most of which would occur during requirements analysis and design—that takes as long to carry out as the typical whole development effort?

Initially, Websites were functionally very simple compared to most traditional software applications, and so the fact that they typically took 8 to 12 weeks to develop, as compared to months or even years for traditional software applications, made some sense. Now, however, Websites and applications have gotten more and more complex, and in many cases are much like traditional applications that happen to be implemented on a browser platform. The industry needs to adapt its notion of reasonable, feasible, and effective time frames (and budgets) for developing complex Web-based applications, which simply are not the same as simple content-only Websites. This includes adapting its notion of what kind of usability engineering techniques should be invested in.

In a report by Forrester Research, Inc. (2000b), called "Scenario Design" (their term for usability engineering), it is pointed out that

Executives Must Buy Into Realistic Development Time Lines and Budgets

The mad Internet rush of the late 1990s produced the slipshod experiences that we see today. As firms move forward, they must shed their misplaced fascination with first-mover advantage in favor of lasting strategies that lean on quality of experience.

- **Even single-channel initiatives will take eight to 12 months.** The time required to conduct field research, interpret the gathered information, and formulate implementation specs for a new web-based application will take four to six months. To prototype, build, and launch the effort will take another four to six months. This period will lengthen as the number of scenarios involved rises.
- **These projects will cost at least $1.5 million in outside help.** Firms will turn to eCommerce integrators and user experience specialists for the hard-to-find-experts, technical expertise, and collaborative methodologies required to conduct Scenario Design. Hiring these outside resources can be costly, with run rates from $150K to $200K per month. This expenditure is in addition to the cost of internal resources, such as project owners responsible for the effort's overall success and IT resources handling integrations with legacy systems.

We agree that 8 to 12 months is a more realistic time frame (than 8 to 12 weeks) to develop a usable Website or application that will provide a decent ROI. Also, if this is the overall project time frame, there is ample time in the overall schedule to carry out a usability engineering program such as the one laid out earlier in this chapter. The total cost of the sample usability engineering program offered here is also well within Forrester's estimate of the costs of "hard-to-find-experts," which include "user experience specialists." The sample cost-justification analyses offered here—as well as others offered by a Forrester report called "Get ROI from Design" (2001)—suggests that it is usually fairly easy to justify a significant time and money investment in usability engineering during the development of Websites, and the framework and examples presented in this chapter should help you demonstrate that this is the case for your Web project.

Portions of this chapter were excerpted or adapted from the book *The Usability Engineering Lifecycle*, Mayhew, 1999, used with permission. Portions of this chapter were also excerpted or adapted from the book *Cost-Justifying Usability*, Bias and Mayhew, 1994, used with permission. Other portions were excerpted or adapted from an article called Investing in Requirements Analysis by Deborah J. Mayhew that first appeared on http://taskz.com/ucd_discount_usability_vs_gurus_indepth.htm in September 2001. Portions of this chapter will appear in a chapter by Deborah J. Mayhew in *The Handbook of Human–Computer Interaction*, Jacko and Sears, in press.

REFERENCES

Bias, R. G., & Mayhew, D. J. (1994). *Cost-justifying usability.* Boston, MA: Academic.

Bias, R. G., Mayhew, D. J., & Upmanyu, D. (in press). Cost justification. In J. Jacko & A. Sears (Eds.), *The handbook of human–computer interaction.* Mahwah, NJ: Lawrence Erlbaum Associates.

Forrester Research. (1998). *The age of Net pragmatism.* Cambridge, MA: Sonderegger, P.

Forrester Research. (1999). *Measuring Web success.* Cambridge, MA: Schmitt, E.

Forrester Research. (2000a). *The best of retail site design.* Cambridge, MA: Souzai R. K.

Forrester Research. (2000b). *Scenario design.* Cambridge, MA: Sonderegger, P.

Forrester Research. (2000c). *Travel data overview.* Cambridge, MA: Harteveldt, H. H.

Forrester Research. (2001). *Get ROI from design.* Cambridge, MA: Souza, R. K.

Faraday, P., & Sutcliffe, A. (1997). Designing effective multimedia presentations. *Proceedings of CHI '97,* 272–278.

Jacko, J., & Sears, A. (Eds.). (2002-in press). *The handbook of human–computer interaction.* Mahwah, NJ: Lawrence Erlbaum Associates.

Karat, C.-M. (1989). Iterative usability testing of a security application. In *Proceedings of the Human Factors Society 33rd Annual Meeting* (pp. 273–277). Human Factors and Ergonomics Society.

Lee, A. Y., & Bowers, A. N. (1997). The effect of multimedia components on learning. *Proceedings of the Human Factors and Ergonomics Society,* 340–344.

Mayhew, D. J. (1999). *The usability engineering lifecycle.* San Francisco: Kaufmann.

Mayhew, D. J. (2001). Investery in Requirements Analysis, www.taskz.com/ucd_invest_req_analysis_indepth.php

Nielsen, J. (1993). *Usability engineering.* Morristown, NJ: Academic.

Whiteside, J., Bennet, J., & Holtzblatt, K. (1988). Usability egineering: Our experience and evolution. In M. Helander (Ed.), *Handbook of human–computer interaction.* Amsterdam, Netherlands, North-Holland.

Wixon, D., & Wilson, C. (1997). The usability engineering framework for product design and evaluation. In M. Helander, T. K. Landauer, & P. Prabhu (Eds.), *Handbook of human–computer interaction* (2nd ed.). Englewood Cliffs, NJ: Prentice-Hall.

5

Web User Interface Development at Oracle: How the Evolution of the Platform Influenced Our Products' Human-Factors Efforts

Luke Kowalski

Oracle Corporation

INTRODUCTION, BUSINESS, AND ORGANIZATIONAL CONTEXT

As the second largest software company in the world, Oracle provides integrated solutions in the realm of applications, servers, and tools. The common perception is that Oracle only sells databases that contain a few graphical user interfaces. The reality is that we have a team of 60 human-factors professionals working on the development of our software across all domains.

In the realm of Applications, the business trend that has affected our user profile and thus the interfaces has been toward self-service. Tasks formerly handled by administrators, assistants, and Information Systems (IS) professionals are now outsourced or delegated to the individual user in the corporation. The browser is the usual delivery vehicle. Some examples of self-serve productivity applications include tools for managing procurement, human resources, manufacturing, customer relationships, and collaboration.

With Servers, the strategy has involved integration and simplification, a process that can lead to improved usability. Some products in this suite serve as the Internet platform or software infrastructure for e-business. Others support the database and its associated tuning, Web monitoring, and business intelligence applications.

The last division, Tools, encompasses products to support developers coding in Java, in native environments such as Windows, and for Web-based UIs. Portals, Java, XML, and other tools fit into this classification.

The efforts to integrate and consolidate most of the products using HTTP-based technology were first made internally. The applications were deployed to our own employees. The company has achieved internal savings as a result of this exercise. Our own human resources, procurement, and other processes that are involved in sustaining a business all became automated. The applications themselves were first hosted on the intranet. The users benefited by having more consistent interfaces, integrated functionality, and ease of access.

To just put software on the Web is not sufficient. A software company has to study the way business is done and tailor applications to support the latest trends. Only after this need is met can it worry about layout, look and feel, and general usability. The trend Oracle is underlining in its development efforts is one of simplicity and completeness, as well as user-centered design. Instead of collecting the "best of breed" applications from several different vendors, we are proposing to create a universal and integrated solution based on what users tell us they need. It will cost less to deploy because it comes ready to use (no professional services needed) and can be hosted by Oracle (removing the need for a dedicated and expensive information-technology infrastructure).

Usability and Design at Oracle, Human Factors in Web Development

The importance of usability has grown with the advent of the Web. Now, user interfaces are more accessible and prevalent, and it no longer takes an expert to put one together. Oracle executives, being fully aware of this, saw the importance of having a strong UI group. The UI and Usability department is a centralized resource at Oracle, now 60-person strong, tasked with supporting the company strategy of integration, ease of use, and simplicity. To have maximum impact, we strive to get involved as early in the product lifecycle as possible. User-centered design starts with gathering requirements from actual users of the software. More often than not, these requirements are assumed or imagined. A technique called Wants and Needs has been refined (customized) by our usability engineers with the purpose of eliciting this information and ensuring that a given functionality is what users in a given domain really need. In a focus group environment, we are able to ask about tasks and objects, get a sense of prioritization by doing a simple card sort, and obtain some sense of functional groupings. Later stages in the user-centered software development cycle include design, architectural integration, usability testing, prototyping, and follow-up with users. As with other design disciplines, all of this happens on an iterative basis.

Throughout this process we strive to create a sense of a virtual team. If we come in at the end of the development cycle as consultants and simply critique the design, we are perceived as an outsider, and few recommendations are implemented. Evolving the design and trying to participate in all stages of development produces much better results, as demonstrated by higher task completion for the products.

Even with 60 people, it is not possible to cover all products. Because of this, the strategy has been to focus on critical products, whereas other integrators tend to follow already established models or the path of least resistance. We call it our policy of "Trickle-down Software Reaganomics." The UI group's focus has lately been on the corporate adoption of standards and tools that create standards-compliant applications. See the "Process changes" section for a detailed description of how this changed our role and, as a result, the usability of our products.

An additional component in our corporate human factors strategy is the new research department. Its role is to stay at the forefront of UI research, with data visualization and mobile (PDAs and wireless) domains being just two areas of interest. Research findings are applied to products at the architectural level, and enhancements are made at the technical or display level as well. The department's charter is expanding to involve active collaboration with top academic human-factors groups and involvement with federal and consortium software standards (Universal Access Board, W3C Web Standards Consortium).

Evolution: How the Web Changed Everything and How the Thin Client Went on a Diet

First generation of our products was implemented with the windows operating system in mind. The second generation transitioned them to the Java platform, or what we originally believed was the "Thin Client." With Java, we gained platform independence, ensuring that the user experience was the same. The technology stack however, did not constitute a "platform" but a loosely tied and primitive set of coding standards. Missing were declarative tools and a mobile strategy. It was a cohesive solution, but it only worked at a superficial level.

Our next generation of Web software has evolved from a Java implementation to a true thin-client model based on HTML, often hosted on the Internet, with Oracle acting as the application service provider (ASP). Outlined next are the differences, process changes, and new factors we had to consider when making this transition.

Technical changes. Software now has to run on the Net, in all browsers, and guarantee universal access (Americans with Disabilities Act of 1990). We no longer have to rewrite the software for every device and delivery vehicle or country. Applications may be hosted on Virtual Private Networks, using Virtual Private Databases. There, customers can safely share a single instance of the application or a database. With this approach, software is also now easier to patch and update. When it comes to suites, the new technology supports single sign-on when the user moves between different applications. The new technology stack itself is based on

XML and JSP architectures. In short, the technical changes no longer hindered the creation of a unified user model and a consistent experience. They served as the groundwork for creating a Graphical User Interface (GUI) layer and for connections between various integrated products.

Process Changes. In designing the third generation applications, we had to adapt to several new business trends. First of all, the development time shortened considerably due to market pressure caused by the proliferation of start-ups. The delivery of the software changed from being shipped in a box to being deployed on the Web (internal or external to the company). This provided us with an ability to change things after the customer started using the software. Teams were addressing software defects, as well as making usability enhancements in weekly cycles instead of the traditional yearly product releases.

Our design team also had to think about new metaphors or blend existing ones. We were now faced with a hybrid of a popular Website model and application functionality. In other words, we had information and tasks put together. Examples of this and how we solved it appear in the case studies below.

One of the most successful tools in our human-factors arsenal are standards. We follow them when it comes to usability testing, using the National Institute for Standards Testing's (NIST's) Common Industry Format and also the Standardized Usability Measurement Index (SUMI) metrics, collected as part of the testing process. But an even more effective tool is our Browser Look and Feel Standard. It is a series of specifications, dealing with UI flows, screen types, look and feel, as well as with individual widgets and controls. This standard is there to ensure a consistent visual and navigational experience for our diverse cross section of users. It is even more important in a scenario where products are integrated as a "do it all" suite. Having all of these things specified in the guidelines is usually not enough in a large organization, as it is hard to educate, police, and enforce in an organization with 30,000 employees. The document itself is also not something that can be processed during a lunch break. What makes the effort effective in our case is a set of tools. This is a code template library and a programming environment to enforce and support standards. Our consultations are still required, but the job of evangelizing and educating is cut in half.

Lessons. In the initial stages, when we transformed existing products to new technologies, we simply translated them literally from Windows to Java and then from Java to HTML. In these retrofits, there were instances when the basic UI model did not need to change, but in most cases, the user interface had to be adapted to a new medium, a new user profile, and a new usage scenario.

One good example of this is in the realm of purchasing. The old UI model we had been using in our applications was the purchase requisition and the purchase order. These are not foreign concepts if you work in a large corporation, but to the "person on the street," the difference is not always clear. With the advent of shopping on the Web and e-commerce, it became popular to use the metaphor of the shopping cart (Amazon, eToys, and Yahoo Shopping). We started using it in our procurement applications. Although we received initial complaints from people in large Fortune 500 companies who were used to the old terminology, we always had excellent task-completion data in the usability lab. There were never any questions about what the cart meant or did, at least not from the end users. Now that the users were ordering the products themselves, without relying on the corporate procurement department, it became more important to meet their expectations and not those of procurement personnel.

We have not solved all problems. Browser UIs pose challenges when it comes to creating information-rich applications that require quick data entry. Web pages do not have stateful screens and have to tediously redraw the entire page or make a round trip to the server in response to simple user actions. Typical master and detail relationship screens and drag and drop behaviors are hard to implement in HTML. Modal windows, a very useful feature in native UIs, have been discouraged for technical and accessibility reasons, often creating long, scrolling pages. Technical users complain of low performance, lack of information density, and too many clicks. We hope to address these with new professional UI standards, and more research as the Web itself evolves.

Case Studies

Unified Messaging. Unified Messaging is a suite of collaboration and communication tools. It includes hosted mail, a calendar, a directory service, and a resource scheduler. All of these were formerly separate products.

As stated previously, the new corporate model has been first to deploy products internally and then to ship them to customers. We took this approach with Unified Messaging. Normally, it would be too expensive to try this, but stability and user satisfaction were deemed more important than "first to market" status. In addition, the entire suite of tools was deployed from a single server. If this had been a traditional application, it would have involved an installation step at every desk in the company.

The guidelines were used as a reference in this project, and the engineering team consulted us if there were questions about interpretation. Because multiple teams across many divisions were involved, the usability team served as a coordinator. The user profile was quite complex, including absolute beginners and power technical users, so it was hard to satisfy both camps. For one group and for one set of functionality, we focused on task completion, for the other, on speed and easily surfacing advanced functions.

The design itself was not revolutionary. It was important to use existing metaphors (Yahoo Mail, Netscape Navigator, and Microsoft Outlook) rather than reinvent the wheel. At the same time, we did attempt to blend Windows and Web metaphors. We wanted transfer of learning from people exposed to both.

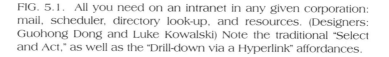

FIG. 5.1. All you need on an intranet in any given corporation: mail, scheduler, directory look-up, and resources. (Designers: Guohong Dong and Luke Kowalski) Note the traditional "Select and Act," as well as the "Drill-down via a Hyperlink" affordances.

The hierarchical structure of the application was not unlike that of a Web store. There were global buttons (round icons on the image above, in the upper right-hand corner of the screen), main level tabs, subtabs, and a persistent search area just below the main navigation widgets (see Fig. 5.1). The functions themselves could be accessed in two ways. For a user used to a traditional desktop application, he or she could select an item by clicking the checkbox (see Fig. 5.1) and then select a button from the button bar to do something (select and act, or noun/verb style behavior). This model is a very efficient path when it comes to actions performed on multiple objects (messages). Conversely, for those whose first GUI experiences were on the Web, we provided a "hyperlink" model (drill-down by clicking "Subject" column). The same functionality (delete and edit) would be surfaced on the screen succeeding this one (the item detail page). This is a good example of accommodating both Web and desktop navigation models. People will eventually chose which one they prefer or never even discover an alternative path.

Unified Messaging was a good example of integration. Several teams came together and made their best attempt to align with the corporate standard. We were able to propose layouts early in the development process. The final result was well

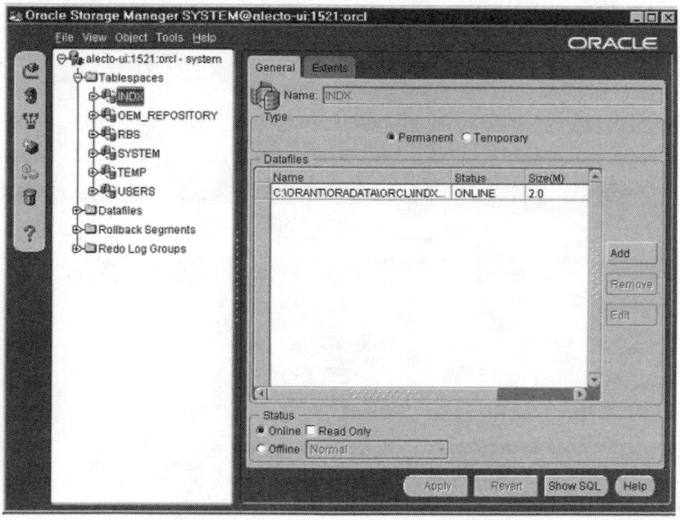

FIG. 5.2.   The old Java version of Enterprise Manager. (Designers: George Hackman, Marilyn Hollinger, and David Abkowitz)

received by users. Although there are still a couple of products in the suite that did not follow the model, this will change in the next release.

Enterprise Manager—A Lightweight Monitoring Framework.   Enterprise Manager is a thin-client framework for administering and monitoring databases, associated services, and hosted applications. It includes a graphical user interface for system administrators and database administrators (DBAs). Status, performance, and some configuration functionality can all be accessed here.

We did not want to translate directly from the Java version of this product (see Fig. 5.2). The old layout included a typical topology tree and detail pane combination (horizontal master/detail layout). The tree was not reused in the new thin-client reinterpretation for four reasons:

- We changed the model because the tree scaled so poorly and caused performance problems in the old product. (Too many nodes created a vertical scrolling problem.) We instead pursued a more task-oriented model.
- Use of frames (for a master-detail-style model) was precluded in our corporate standards because of accessibility, performance, and security reasons (screen readers have a difficult time with frames)
- The tree without a frame would cause a full-page redraw on every click, and this was considered unacceptable.
- Oracle has a requirement to support older (and multiple) versions of certain browsers that presented further problems.

FIG. 5.3.   Console view of Enterprise Manager. A grouping of commonly accessed information, completely customizable, depending on job role, functionality accessed, or the corporate security model. (Designers: Michelle Bacigalupi and Luke Kowalski)

What we chose instead was a progressive disclosure model. As the users drill down the hierarchy, we would surface a trail of hyperlinked "breadcrumbs," allowing them to retain context. Each time an object in a given table was opened, the user saw a new context and the possibility for further drill-downs (see Fig. 5.5). The main entry page (see Fig. 5.3) is an amalgamation of the most commonly accessed functionality, arranged in portlets. These shortcuts provide an overview and a general system status at a glance. It is customizable for each installation and for each user.

Enterprise Manager, or the Internet Application Server (IAS) monitoring framework, as it came to be called, was a prime example of a new type of collaboration where the process improved the usability of the software. Positioning ourselves as the virtual team members helped. Early inquiry into what users wanted helped us aim the product better before design actually started. Recommendations from several iterative testing sessions improved on our assumptions. The fact that we did not just copy what was in Java into HTML made for an improved product as well. Our specifications and designs were posted on an internal server, where

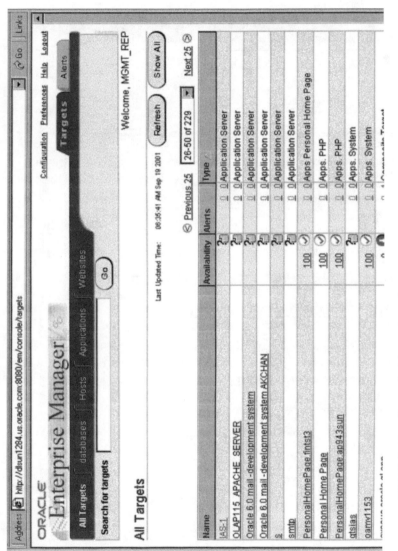

FIG. 5.4. A roll-up of objects the administrator can manage. We got here by clicking one of the links in the portlets above. (Designers: Michelle Bacigalupi and Luke Kowalski)

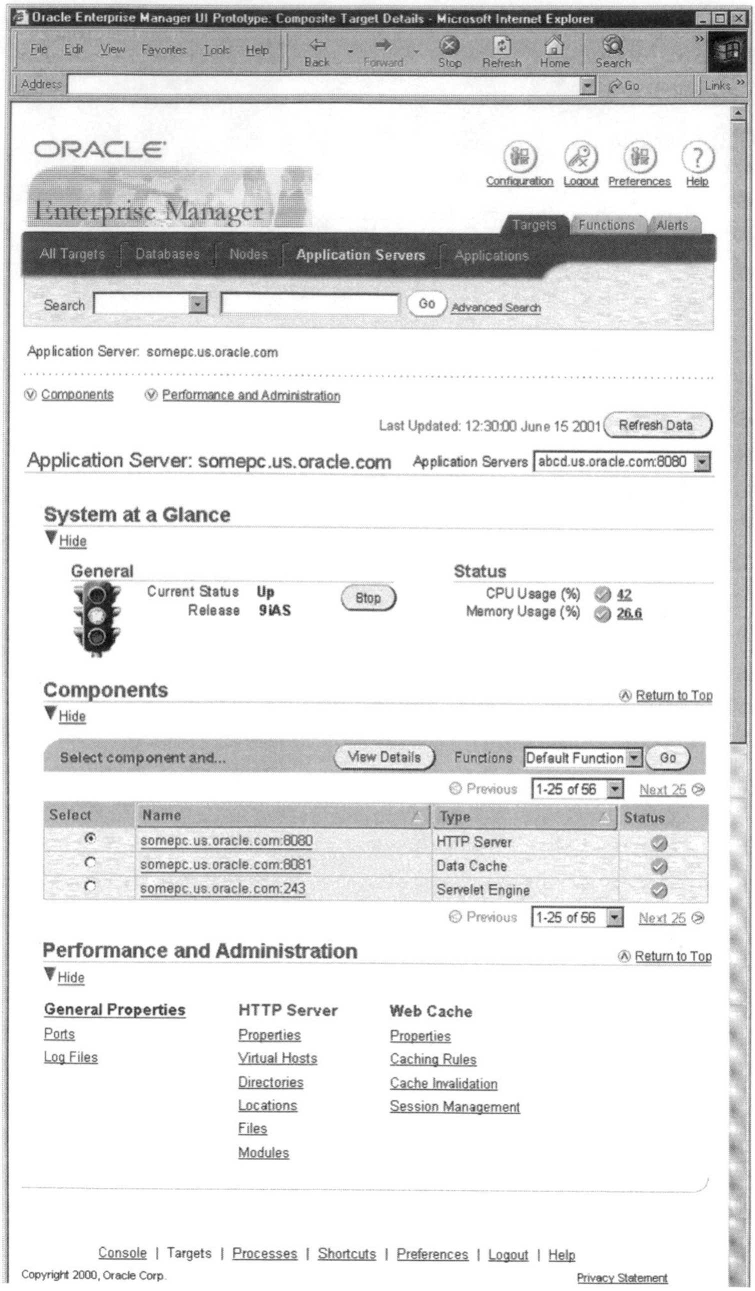

FIG. 5.5. A drill-down into a particular object. This page contains information about its status, common operations, and links to related information or to its children. We chose anchors as a way to deal with long vertical pages and the high density of information on a page. (Designers: Michelle Bacigalupi and Kristin Desmond)

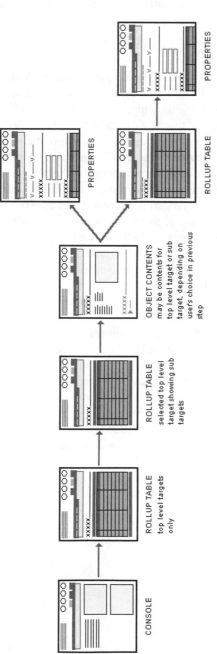

FIG. 5.6. Schematic for navigation. An example of how to work around a tree in a browser-based UI by navigating through hierarchies of master and detail structures. Three separate prototype walk-throughs were conducted to validate this model for the navigation. (Schematic authored by Kristin Desmond and Michelle Bacigalupi)

they were used as the living specifications, helping integrators around the globe interpret the designs. We followed this up with design reviews, further ensuring quality, consistency, and alignment with the administrative and general corporate look and feel guidelines.

Future and Conclusion

Web application development at Oracle has accelerated in the last 3 years. The importance of human factors gained considerable visibility in the executive suite. As part of the corporate mandate, the company strategy evolved to include the direction to simplify, integrate, and improve ease of use. We have gained new tools in our human-factors work by learning to engage users before the software architecture is frozen. The corporate standards for HTML interfaces have gone a long way in ensuring simplicity, integration, and consistency. By themselves, they tend to be thick illustrated volumes that nobody ever reads. But coupled with programming tools to automatically generate standards-compliant user interfaces, they became real. We believe that it is because of these efforts that we are already seeing improvements in task completion rates, as well as favorable reviews in the press.

Not all applications at Oracle were transferred to the Web. There were some cases in which it did not make sense to do so. Software that involves intense data entry or programming tools is not appropriate in a browser. The need for speed, what you see is what you get (WYSIWYG), and nonlinear processing outweighed the benefits offered by the Web.

The next step is to work on the third generation tools that would include stateful UIs (no page redraws) while monitoring how HTML standards evolve to allow for more interactivity, pop-up screens, and other elements. Once again, when we get there, it will be important to reexamine the functionality of each application and suite. We will once again need to decide whether it can be translated as is or whether it needs to be redesigned to adapt to the new medium and, in effect, the next generation business practices. We will always test it with the users to see if we were right or wrong, and the marketplace will provide the final answer.

6

Usability Metrics

Robert Opaluch
Verizon Labs

WHAT ARE METRICS?

A metric is a repeated quantitative measurement that is centrally important to an effort (e.g., providing a Website). When working on an important project, it is useful to have quantitative measurements that capture the success or signal problems in the effort.

Metrics Are Quantitative

Keeping track of top usability problems or doing Web design might be very important for a project but by themselves are qualitative not quantitative. These important activities, therefore, are not metrics.

Usually, it is easy to turn qualitative data into quantitative measures in one way or another. For example, if keeping track of top usability problems are important for a project, count the number of outstanding usability problems by severity and track these numbers over time. If a user's quick completion of an ordering task is an important criterion in design efforts, track the time to completion of that ordering task. Qualitative information can be transformed into quantitative metrics.

Metrics Are Measurements

A set of metrics must be valid, reliable, and complete. Measurement issues need to be addressed whenever metrics are employed. This chapter can't begin to summarize these issues in detail, but plenty of references are available on measurement.

Validity

Validity refers to measuring what was intended to be measured. To measure whether users can complete a product ordering task successfully, the percent of users completing that task within a reasonable period of time in a usability test appears to be a valid measure.

Reliability

Reliability means that different measurements of the same thing provide consistent results. For example, asking two similar questions in a questionnaire should result in a very similar or identical rating. A measure must be reliable before it can be valid.

Metrics Are Important Indicators

Especially when doing server log analysis, there is a lot of quantitative data that is relevant to usability efforts and is worth examining. What are the project's real purposes, goals, objectives, and targets? What important numbers need to be tracked in order to understand if the project is succeeding in meeting those targets? These most important numbers should be the metrics that guide the decisions and form the evaluation of how the project is doing.

For example, if the purpose of a Website is cost savings and revenue generation, the metrics might include online revenues and the number of online transactions (which reduce the costs of call center volumes). Related usability metrics might be the percent of users who could successfully complete online purchases and cost-saving transactions. These measures are centrally important to the success of the Website effort. Other numbers, such as page views or number of visitors, might be numbers worth knowing and examining, but they are not the metrics that ultimately determine the success of the effort. On the other hand, if the purpose of the site is to sell advertising and maximize ad revenues, the number of unique visitors might be a centrally important number that determines the success of selling advertising and charging higher ad rates.

Analyze, ponder, and agonize over what really matters in measuring the success of an effort. The numbers that seem important initially may not be fundamentally important to the success of the effort. Focus on tracking numbers that are centrally important to success.

Be Careful When Selecting Metrics

Sometimes, selecting metrics may have consequences not anticipated when the selection is made. Often, just the act of measurement changes how people behave. Telling people that they will be evaluated or compensated based on certain metrics is likely to affect their behavior.

For example, telling employees that they will be evaluated based on customer service ratings from a questionnaire is likely to change the behavior of those employees toward their customers. The intended effect of the questionnaire is to improve employees' service to customers. However, an unintended effect may surface as well. For example, an employee might cajole or press customers to give better evaluations because it would affect that employee's income. The manager that selected the customer rating questionnaire metric probably did not anticipate that employees would ask customers for better ratings but that they would earn better ratings through better service.

Consider how selection of a usability metric might affect the behavior of designers, testers, and others. For example, if the proportion of users completing an ordering task is the metric that a designer is trying to optimize, one might expect the designer to bias their designs toward simple, straightforward designs that would allow users to complete the order. Unintended effects might include making the ordering task easier to complete at the expense of other goals, such as meeting user interface design standards or aesthetic considerations, or providing ordering options that were intended to be available to users. Metrics may have to become more detailed and specifically mention unintended effects in order to eliminate those effects when the data is analyzed.

Wilson noted that testers can be reluctant to report a problem when that problem might affect the team bonus (Molich, Laurel, Snyder, Quesenbery, & Wilson, 2001). The selection of metrics that do not result in unintended side effects requires not only proper analysis of goals, targets, and objectives, but also experience and a willingness to make modifications. Be careful what you wish for; be careful what metrics you select.

Metrics for the Web

Metrics are important for Website design and analysis. Examples of Web metrics include

- Usability metrics—ease of use, or user friendliness, of the site.
- Performance metrics—site availability and load.
- Traffic metrics—site visits, page views, and download times.
- Business metrics—revenues, cost savings, and return on investment.
- Call Center metrics—transactions, staff efficiency.
- Customer-centered metrics—cycle times, service-level agreements, and customer service metrics.

- Competitive analysis—direct competitors, other alternatives, and best-in-class comparisons.
- Developer report card metrics for evaluation and bonuses.
- Process quality metrics—customer satisfaction, cycle times, defects, and costs.

WHY USE METRICS IN WEBSITE DESIGN?

Change Discussions from "Subjective Opinions" to "Objective Hard Data"

Project staff often argue about usability. Is one design more usable than another, or is a redesign better? Project team members may claim that one design is easier to use, an improvement over alternatives, or that some design alternative poses a usability problem, whereas others may have differing opinions. Changing discussions from qualitative to quantitative can result in putting issues in perspective. In quantitative analyses, we need to count or gather data to support claims about usability and be more precise in evaluations.

Select Metrics to Monitor Progress Toward Goals

Setting goals often involves collecting metrics to see if goals are met. Most usability goals tend to be quantitative. A good reason to collect metrics is to allow early and precise detection of usability problems and failure to meet goals, objectives, and targets.

See Progress Over Time

Although time charts can be constructed that show qualitative data, most graphs plot a quantity over time. By looking at quantitative measures, it is easier to detect improvements and deterioration and to detect it more quickly than when using qualitative measures only. Usability engineers can track the number of usability problems identified and resolved over time. Problems can be ranked by severity, with different severities tracked separately or combined into weighted averages. The impact of usability problems on revenues, cost savings, or other measures can be calculated and tracked over time as well.

Make Precise Comparisons With Other Sites and Competitors

User performance (e.g., successful task completion, time on task, and number of errors) can be tracked on key tasks and compared to performance on competitor sites, earlier versions of a site, or alternative site and page designs. Comparisons can also be made among methods for accomplishing the same task without using the Web (e.g., brick-and-mortar store purchases).

Make Informed Decisions

Decisions can be made on numbers rather than less detailed analyses. For example, rather than make decisions based on opinions about which design appears to be easier or better in some way, user performance can be observed and differences calculated.

Show the Impact of Usability Problems on Business

Revenue, cost savings, and traffic on Websites are all affected by usability problems. These effects can be calculated and tracked. The results can be used to justify and prioritize usability improvements to the site. See Bias and Mayhew (1994) for more analysis of cost-justifying usability.

Make Financial Estimates

The impact of usability problems on cost savings or revenues can be calculated and tracked. For example, usability problems may lead to failure to complete an online purchase. The proportion of users failing the task can be used to estimate financial losses due to failure to sell products and services on the Website. The costs of users calling a representative to complete a purchase versus the Web can be used to estimate cost savings. Return on investment calculations can be done to estimate whether a change in a Website is financially justified by the cost savings or increase in revenue expected from the change.

Size Hardware and Software Appropriately

Web traffic measurements can be used to determine the infrastructure for the site. Web response times are important to users and usability engineers, as well as those responsible for hardware and software procurement and administration.

Meet Contractual Obligations

Businesses may have contractual obligations to offer service within specified time periods. Service-level agreements can be offered to customers and tracked quantitatively.

Monitor Customer Satisfaction Levels More Precisely

Customer satisfaction is typically measured and tracked quantitatively over time for greater precision and to detect customer service problems early.

Track, Manage, and Improve Usability Engineering Methods

Using the right methods can greatly increase the chance of successful outcomes. If usability engineering best practices are performed, the product or service being designed is likely to be more usable than those designed without adherence to good usability engineering practices. Methods can be measured by tracking process checklists. A project can track the cost of methods, and the number of usability defects which are identified and fixed as a result. The cost savings of early identification of problems can be calculated and tracked. The cost of fixing usability problems after deployment is often 10 to 100 times more expensive (Mayhew, 1999). Web pages or entire sites are often redesigned to fix usability problems. Projects and companies can track which methods are most effective at detecting usability problems and fixing them early.

Impress Executives

Executives like to run their companies and organizations by numbers. Give them the metrics they need to make the right decisions. Appeals to management are more likely to be successful when they are backed up by metrics.

COMMON SOURCES OF USABILITY METRICS DATA

Usability Questionnaires

User opinions on different aspects of usability often need to be collected to measure customer or user satisfaction and acceptance. Standard usability questionnaires include the Questionnaire for User Interaction Satisfaction (QUIS), available for licensing from the University of Maryland; the Computer Usability Questionnaire (CSUQ), available from IBM (Lewis, 1995); and the System Usability Scale (SUS; Brooke, 1986). See Fig. 6.1 for an example of some usability questions from a questionnaire.

Typically, usability questionnaires employ questions that ask at least 5 different scale points and as many as 10. Scales may be unbalanced (from none to very high on some ratings) but more often are balanced scales (from very negative, to neutral, to very positive on some ratings).

Please circle the number that best captures your opinion about each of the following statements.

1 = Strongly agree, 2 = Agree, 3 = Somewhat agree, 4 = Neutral,

5 = Somewhat disagree, 6 = Disagree, 7 = Strongly disagree

1.	This Website was easy to use.	1	2	3	4	5	6	7
2.	I could adjust or customize this Website to my preferred way of working.	1	2	3	4	5	6	7
3.	Icons, graphics, charts and diagrams on this Website were helpful.	1	2	3	4	5	6	7
4.	It was easy to do the account set up, subscription, or downloads before using this Website.	1	2	3	4	5	6	7
5.	I was satisfied with the customer service call center associated with this Website.	1	2	3	4	5	6	7

6.   What did you like best about this Website?

7.   What did you like least about this Website?

8.   How could we improve this Website?

FIG. 6.1. Usability Questionnaire example questions.

Scale points may be labeled only at the end points or at all of the points on the scale. For example, labeling only end points would rate "ease of use" from 1 to 7, where 1 = very easy and 7 = very difficult. Labeling all points would have respondent's rate "ease of use" as

1. Very easy.
2. Easy.
3. Somewhat easy.
4. Neither easy nor difficult.
5. Somewhat difficult.
6. Difficult.
7. Very difficult.

The latter is better for reporting descriptive data (e.g., 62% of users rated the system as easy to use or very easy to use), the former for inferential statistics or research relating responses to one question to other data, such as user background or observed behavior.

Usability questionnaires ask questions about overall ease of use; different aspects of ease of use, such as ease of learning, efficiency, error rates and recovering from errors; different aspects of the evaluated system or service, such as consistency, organization, navigation, text messages, and graphics; and impressions of the system, such as frustrating, stimulating, and comfortable. Sometimes usability questionnaires also ask about the functionality, features, and value of the system, as well as its usability. They also may contain items that collect demographic data on the background and experience of the user filling out the questionnaire.

Online Usability Questionnaires

Although online usability questionnaires are being used frequently, they may be problematic as a main source of user opinion data. Online questionnaires suffer from a sampling problem: Only a small fraction of users choose fill them out, so they are a self-selected group, not a representative sample. Online questionnaires can also be intrusive and have a negative impact on the user's experience with a site. However, online questionnaires can be designed to be less intrusive and can provide a useful adjunct to data collected elsewhere. With a large number of responses, online questionnaire data is likely to provide some useful feedback that is not identified elsewhere. If an online questionnaire is used, consider the following recommendations:

- Keep it short. Typically, a questionnaire should only pose a few questions to the user. Use longer questionnaires only for selected users in special programs such as pilot programs and beta tests.
- Avoid presenting it in the middle of a task, as an interruption can have a disruptive effect on the user (Bailey, Konstan, & Carlis, 2001).
- Time the presentation of the questionnaire to be minimally intrusive, such as
  - At the completion of important tasks (e.g., after an order is completed).
  - At the completion of important steps (e.g., after search results are presented).
  - When the user requests help.
  - When an error has occurred.
  - When a message is presented to the user.
  - When the user attempts to leave the site.
  - When the user clicks a voluntary link, such as "Rate this . . . "
- Keep it focused. Consider online questionnaires targeting a particular feature or sequence, not a general feedback questionnaire.

- Target the content of the questionnaire on the immediately preceding activity of the user. Questions can be specific to the page being viewed by the user.
- Open-ended questions might be preferable to multiple-choice questions, because the data gathered is probably better suited to getting new ideas, not calculating user satisfaction levels in general.
- Focus on functionality questions, not too much on usability evaluation. Questions on usability might be limited to a single rating scale on ease of use or satisfaction, with other questions on functionality.
- Provide an incentive. Incentives might help increase response rates.
- Ensure adequate response rates. Low response rates may invalidate any conclusions drawn from the questionnaire. Consider this risk when using rating scales. Be sure to describe this risk and limit conclusions when analyzing and communicating the results on online questionnaire data.
- Be cautious about interpreting online questionnaire data. Short questionnaires often cannot include questions that would help remove questionnaire responses that are invalid due to users not really reading or providing appropriate and honest answers. Longer questionnaires can include items that detect invalid responses. Keep this in mind when evaluating online questionnaire data.
- Consider ways of getting data besides use of an online questionnaire. For example, click to chat or e-mail links might be more appropriate to solve some problems than would a questionnaire.

When providing a link to a questionnaire that is always available to the user, consider these recommendations:

- Feedback link is always visible.
- A link is always available, on every page.
- A link is located at the same place on every page.
- Keep the current page visible.
- Provide quick rating scales and optional open-ended data entry.

User and Customer Interviews

Interviews are structured question-and-answer sessions conducted with customers and users. Interviews are useful for collecting new ideas in a wide range of areas but do not typically generate much quantitative data. However, interview questions can ask respondents to select one of a number of alternatives in an ordinal scale or ask quantitative questions. For example:

1. Compared to other online business management sites, would you say the site is

     a. Much better.
     b. Somewhat better.
     c. Neither better nor worse.
     d. Somewhat worse.
     e. Much worse.
  2. How often do you use the site? How many times per week is that?

In general, tracking usability metrics is better done by questionnaires than interviews, although exploring and analyzing usability issues is better done by interviews and behavioral observations.

Observations of User Behavior

A user's opinion does not necessarily reflect that user's behavior. Therefore, it is wise to collect behavioral data to supplement opinion data. Examples of behavioral measures that could be used as metrics are listed on the right-hand side of Table 6.1. On the left side are comparable opinion measures.

The two main sources of behavioral data for usability metrics include usability testing observations and server log data. Search attempts, "contact us" forms,

TABLE 6.1

Comparing User Opinions and Behavior

Opinion Metrics	Comparable Behavioral Metrics
Overall ease of use	Workload
	Rate of successful completion of key tasks
Ease of setup	Percent of successful downloads, registrations, subscriptions, sign-ins
Ease of learning	Time to learn to do a task
Efficiency and productivity	Time to complete a task
	Number of commands/keystrokes/mouse clicks used to complete a task
	Number of steps to complete task
Error rates	Percentage of tasks completed without errors
	Number of errors per task
	Type of errors per task
	Percent of time spend on errors
	Percent repeated or failed attempts
Recovery from errors	Percent of errors corrected successfully
Customizability	Percent of users able to customize the site or tasks to their needs
Emotional reactions	Ratio of positive to negative comments
	Number of expressions of frustration, anger, negative affect
Customer support	Percent able to complete a customer support transactions successfully
	Percent who reference customer support options
Degree of satisfaction	Percent selecting the Website over alternatives
Effectiveness	Percentage of tasks completed
	Ratio of successes to failures
	Percent of goals achieved

e-mail, data from online forms, other Web data and observational studies of users can be analyzed quantitatively as well. There are many ways to code data that will result in quantitative data that can be tracked over time.

Behavioral Observations From Usability Testing Sessions

Usability testing is evaluation of a user interface design by observing (typically by videotaping) users who fit the typical user profile as they perform key tasks using the prototype design. Usability is conducted for a number of reasons, including:

- Identification of usability problems to fix before product release.
- Compare user performance on the prototype to performance with alternative designs, previous releases, competitors, or alternatives to the Web.
- Determine if usability objectives (usability goals) are met.

For further information on usability testing and other usability engineering methods, consult Mayhew (1999).

Behavioral Observations From Server Log Data

Server logs collect large volumes of data on activity of Web servers. This data can be analyzed for many purposes, including identifying user behavior on the Website. Although there are many problems in correctly identifying user behavior patterns, server log data analysis can provide unique insights into the actual behavior of large numbers of users on an established Website.

Analytic Measurements

Besides tracking actual user behavior and opinions, there are other valid indicators of usability. Judgments of usability experts, conformity to usability design checklists or usability style standards and guidelines, and other usability methods and resources can be tracked in order to assess the likely usability of a product or service under development. Heuristic evaluations (usability assessments by a team of experts), usability design checklists, lists of deviations from style guidelines (or checklists on conformity to guidelines), and conformity to usability engineering methods are data worth tracking. A project team may decide that these analytic measurements may be the some of the best measures to predict a successful product design and select one or more of these analytic measurements as a usability metric for the project. For example, a project team may track the percentage of compliance to an internal checklist or the number of serious outstanding usability problems from heuristic evaluations.

For more information on heuristic evaluation, design checklists, usability style guides, and usability engineering methods, consult Mayhew (1999).

USABILITY OBJECTIVES

Usability objectives are concrete, quantitative, user–system performance goals. Usability objectives specify that a certain type of user can perform a specific task within some quantitative range. Usability objectives can be used in the user interface design process to guide user interface design decisions, to focus user interface reviews, and to guide usability testing (user–system testing). Rather than wait until a system is deployed and then see if users are satisfied with it, goals can be agreed on in advance. System designs can be tested before deployment to determine if the system will satisfy user requirements.

Usability objectives can emphasize a number of different types of usability criteria, which could focus on any number of major aspects of the user experience, including but not limited to

- Initial setup, installation, downloads, and subscription.
- Learning to use.
- Efficiency and productivity.
- Error rates and recovering from errors.
- Emotional reactions, frustration, and enjoyment.
- Working with others effectively.
- Adaptability—customization, personalization, internationalization, and universal design.
- Customer service—manuals, help lines, training, and other support.
- Performance under stress and nonoptimal conditions.

Some examples, including user efficiency and productivity, user error rates, cooperative work, and teamwork are described later. Under each of these three criteria are a number of possible usability objectives.

These sample usability objectives need to be edited before being adopted. A project should adopt only a handful of the most important and most easily measured objectives. In future releases, a larger number of objectives could be set as the objectives become incorporated in the user interface requirements process and user interface design process used by the project.

User Efficiency and Productivity

Users need to complete tasks efficiently and productively. This means

- Completing tasks within the shortest amount of time.
- Doing tasks faster than they are done today.

- Completing as many tasks as possible within each day's work.
- Completing work on a timely basis.
- Being able to take over the work in progress of other users and complete it as quickly as possible.

Examples

We might want to estimate numbers for what would delight, satisfy, and dissatisfy Website users:

- At least 90% of the time, users can create a new order for Product A within 8 minutes.
- At least 90% of the time, users can complete and then create a new order for Service B within 10 minutes.
- At least 90% of the time, users can check order status within 1 minute, starting from the home page.
- At least 90% of the time, users can complete the search orders task within 5 minutes.

User Error Rate

Users need to be able to complete their work with a minimal number of errors being made. This means that users need to be guaranteed that the system will help them minimize

- Unintentional typing mistakes.
- Input of incorrect choices or data.
- Misunderstandings or mistakes when taking over someone else's work.
- Other actions that lead to rework.

Users need to be able to recognize that mistakes were made and fix them easily.

SERVER LOG ANALYSIS

A server log file is a historical summary of events of interest on a Web server. Each line (or record) in the server log includes a time stamp and documents information about server activity. Typically, server log records include date, time, requests for pages, information about the requesting machine, and other events of interest. By placing a cookie on a user's machine or through user registration, users can be tracked through sessions and across sessions.

Analysis of log files is conducted to evaluate server load and traffic. Server logs tend to accumulate large amounts of raw data quickly. Analysis of the collected data can be very tedious, and therefore a server log analysis tool should be used to help organize and summarize that data. Typical reports generated by server log analysis programs include information about Website traffic over time, counts and percentages for pages accessed, number of pages accessed at different times of day and days of the week, errors detected, and identification of visitors. More sophisticated programs can plot diagrams showing common navigation paths through the site, average time spent on pages, and common site entry and exit points.

Log files can be analyzed to examine user behavior, including pages visited most often, patterns of navigation through the site, and success (or failure) of completing tasks. Note that a usability test is effective at identifying high-level task problems, whereas server log analysis is effective at identifying how prevalent those problems are.

Log file analysis is a useful adjunct to other types of usability evaluation, serving to raise new questions and add supplemental data to support usability analysis claims. The details of a server log analysis are often too low-level to gain an understanding of the higher level task being performed. Compared to server log analysis, a usability test produces different, but often complementary, results.

Advantages to Examining Log Files

These include

- Analysis of actual user behavior: What is being analyzed is user behavior, not opinions from users or conjecture from third parties.
- External validity: The behavior being analyzed is occurring in the "real-world" natural environment, not an artificial lab setting, so conclusions can be generalized to the real world.
- Large sample size: Log files often include thousands of users over months, not just a handful of users at one point in time.

Disadvantages to Analysis of Log File Data

These include

- Incomplete records of user events—because the server log records only events at the server, anything that the user does that doesn't send a message to the remote Web server is not recorded. Even requests for Web pages may not be recorded, because a recent copy of the Web page may already reside in a cache on the user's machine or on a local server, and the page is sent to the user without going back to the original Website server to get the page.
- Incomplete information about navigation sequences: Because some of the pages requested by users are not noted in the server logs (as described in the previous point), inferences about page navigation sequences may be incorrect.

- Incomplete or incorrect identification of the user: Server logs contain information about the machine making a request. However, the request may be from an intermediary (e.g., a corporate Internet gateway machine or Internet service provider temporary address) rather than the end user's machine. Even if the user's machine is identified, someone else may be using that user's machine or account, or the same user may use multiple machines. Therefore, information on user identities may be incomplete or corrupted with data from others.

- Incorrect information about time spent on a page: User requests for pages are recorded, not their actual page viewing behavior. If users are distracted or conducting multiple simultaneous tasks, this information is not available to the server. All that is recorded is the amount of time between page requests. Use of the "Back" button is often not recorded because a request may not be made to the Web server. Also, it may take some time to display a page after the request, more for some pages than others, and page viewing times should be considered approximate.

- Difficulty tracking dynamically generated pages: Tracking page statistics, user navigation, or user success at tasks becomes complex when page content is generated dynamically (e.g., using active server pages or Java server pages to produce product pages matched to the user's specific interests). Obtaining useful log file analysis data is still possible (e.g., use of single pixel graphics, which can create a request for a named graphic file in the log file to track specific events).

CREATING COMPOSITE METRICS AND DRILL-DOWN HIERARCHIES

Sometimes a single measure of user opinions or behavior alone does not adequately capture enough information to be used as a usability metric. Data may need to be combined from a number of measures to form one or more overall usability metrics. One way to create composite scores is to collect data and then combine separate scores with or without multiplying each of the separate scores by some weighting. Using weights can emphasize the data that is more important and deemphasize less important data when creating a composite score. Goals or objectives can be set, and deviations from goals can be tracked as well.

See Table 6.2 for examples of composite metrics based upon behavioral data from a number of separate tasks. Note that for computing a composite score for ordering, the more important tasks (e.g., submitting an order) are multiplied by a larger weight (0.30) than less important tasks, such as online requests for a new account number (0.05). In this example, an online order request is weighted six times as much as requesting a new account number when the composite score is computed ($0.30 = 6 \times 0.05$). Note that in this example composite scores for

TABLE 6.2
Building Composite Score Metrics

Metric Type	Metric	Computation				
			Goal			
			Score	Weight	S × W	S–G
Cost savings	**Total Cost Saving Score:**	58%	64%	1.00		0.33
	Online ordering	72%	79%	0.30	0.2379	0.07
	Online repair	72%	77%	0.30	0.2298	0.05
	Online billing	50%	50%	0.30	0.5000	0.00
	Online service record	80%	93%	0.05	0.0465	0.13
	Online product catalog	25%	62%	0.04	0.0248	0.37
	Online learning center	50%	21%	0.01	0.0021	(0.29)
Cost savings	**Ordering Score:**	72%	79%	1.00		0.50

(Continued)

ordering and other functions are then combined to form a composite score for cost savings overall. If these composite metrics were put into a Website, the composite overall score could appear on the site, with the ability to drill down to view specific components that make up the composite score (e.g., the ordering composite score or the repair composite score). The Website could provide the ability to continue to drill down on the ordering composite score to view the components of the ordering composite score (e.g., submitting orders or requesting new account numbers). Again, the ability to drill down into order submissions could be provided to see what types of orders were submitted (e.g., by product or product type). So metrics

TABLE 6.2
(Continued)

Online ordering						
	Submit online order requests	50%	27%	0.30	0.0810	(0.23)
	Online order status	80%	85%	0.20	0.1700	0.05
	Online order public log use	75%	90%	0.10	0.0900	0.15
	New account numbers online	75%	98%	0.05	0.0490	0.23
	Access order request details	75%	85%	0.15	0.1275	0.10
	Access order details	75%	85%	0.15	0.1275	0.10
	Track imported orders	75%	85%	0.05	0.0425	0.10
Cost savings	**Repair Score:**	72%	77%	1.00		0.23
Online repair	Submit online trouble reports	50%	27%	0.30	0.0810	(0.23)
	Online trouble status tracking	80%	91%	0.35	0.3185	0.11
	View trouble reports	80%	82%	0.20	0.1640	0.02
	Line testing	75%	98%	0.05	0.0490	0.23
	Access repair history	75%	85%	0.10	0.0850	0.10
Cost savings	**Billing Score:**	50%	50%	1.00		0.50

can serve not only as a summary or overall indicator of success, but as a hierarchy of information that is most important to the success of the project.

COMPUTATION AND PRESENTATION ISSUES

Metrics are often best communicated visually in charts, graphs, and other visual data displays. The most appropriate choice of type of visual display depends on many factors. Visual displays should be readable and understandable. Overloading a display of data can reduce comprehension by those who are less familiar with the metrics data being displayed. Different types of graphs and charts are meant to emphasize different aspects of data. Select the type of graph or chart that emphasizes the points needing to be conveyed. For example, pie charts are best for emphasizing the proportion of different factors contributing to the whole. Line charts are best for noting trends.

The Web can enable interactive or drill-down presentation of results. Some projects or companies have set up metrics "dashboards," or summaries of progress. Consider organizing metrics into a Website and drill-down hierarchies. Metrics computation issues can become more than simple operations. Have a statistician participate or review metrics computations and data displays to assure that the data is being computed and displayed appropriately.

EMPLOYING METRICS TO IMPROVE PRODUCT USABILITY

In the earlier section, "Why Use Metrics in Website Design?" there were a myriad of reasons cited for using metrics. Keep these reasons in mind when metrics are used on a project. Are you accomplishing what you set out to do in a more precise and positive way using metrics? Below are some steps that might help construct a metrics program. Tailor the program to the reasons for employing metrics.

Understand and Clarify Purposes and Goals

The first and most important step is to understand and clarify what is most central to the success of an effort, as discussed in the earlier section, "Metrics Are Important Indicators." This crucial step is often overlooked. Avoid letting the easy availability of certain server log analysis tool reports to overly influence the decision about which metrics are used on a project. Be purposeful and thoughtful in selecting the most important goals for a project or effort and then deciding on metrics that best measure that success or failure. Identify the measurements that are most central to the success of the product or service being evaluated.

The business metrics chosen might influence the usability metrics to choose. For example, if the purpose of a Website is to drive online revenues, then usability problems associated with tasks that generate online revenues seem important to track. If cost savings or online ad revenues are the business metrics, then usability metrics associated with cost savings or traffic might be the most important to track.

If the purpose of a Website redesign effort is to improve usability or address some specific usability issues, then metrics need to be selected to track progress on resolving the usability problems and promoting the user behaviors or opinions that drove the redesign effort.

Setting usability objectives often leads to sensible choices of usability metrics. Customers and users are the focus in the initial step of setting usability objectives and selecting the type of usability criteria and the degree to which these criteria must be met to meet customer expectations about usability.

Analyze Potential Metrics and Available Measurements

With a clear understanding of purposes and goals, select measures that are available and that would be appropriate metrics, or that would contribute to composite metrics made up from a number of separate measures. Keep in mind the reasons for collecting metrics in the first place. Are the measures under consideration the best way to understand progress toward goals? Are there better alternatives available?

Analyze the pitfalls of the metrics that are selected. How could use of those metrics backfire? What unintended negative consequences might occur due to selecting those metrics?

Also consider the gaps that might not be covered by the selected metrics. Could the metrics turn out meeting goals while other usability problems (not measured) compromise the success of the effort? Could the metrics be broadened or could additional metrics be added to address these gaps? Early and precise detection of usability problems and failure to meet objectives is useful.

Set Quantitative Targets, Objectives, or Goals

Metrics data must be evaluated against targets, objectives, and goals. Consider making comparisons with alternative Websites (or products or services) when setting targets, including

- Competitor's offerings.
- Earlier versions.

- Alternative designs.
- Alternatives to the product or service (including non-Web alternatives).

A target can be set to

- Satisfaction ratings, such as responses to a questionnaire item: "Overall, I was satisfied with this site."
  - Goal: 10% or more very satisfied
  - Goal: 50% or more satisfied or very satisfied
  - Goal: 10% or less dissatisfied or very dissatisfied
- Ease of use ratings, such as responses to a questionnaire item: "Overall Ease of Use"
  - Goal: 10% or more rate "very easy to use"
  - Goal: 50% or more rate "easy to use" or "very easy to use"
  - Goal: 10% or less rate the site as somewhat or difficult or very difficult to use.
- Behavioral data, discussed in depth in the usability objectives section of this chapter.
- Top usability problems, such as
  - No identified usability problem reduces revenues or cost savings by more than 10%.
- Search metrics, such as
  - Goal: 90% of the top 100 search terms return results that are clicked by users 90% of the time or more.
  - Goal: 90% of the top 10 search failures are fixed within 30 days to return results that are clicked by users 50% of the time or more.
- Online help metrics.

Tools, Data Collection and Data Analysis Procedures

The amount of time and investment in procuring, setting up, and administering metrics tools may be the greatest cost of a metrics effort for complex commercial Websites. Data collection and data analysis often requires help from those with expertise in questionnaire design, data analysis and statistics, usability testing, and behavioral psychology. Be sure to rely on those with the right background and experience to assure that the data collection effort is objective, accurate, and as effective as possible. It is easy to create questionnaires, do behavioral observation, and analyze data, but it can be difficult and time-consuming to do it well. Those with little experience and training may collect data that is biased, unintentionally inaccurate, or minimally analyzed.

Evaluate Metrics Results

When data is compared to targets and analyzed in detail, there are likely to be many opportunities for improvement, including:

- Improving the Website, product or service being evaluated.
- Changing the direction or emphasis of research, design, or development because of discoveries made while analyzing metrics and other data.
- Improving communication and understanding of goals, targets, or objectives and how well they are being met.
- Adjusting or improving the metrics themselves, or adding additional metrics.

Implement Improvements Based on Metrics Results

When metrics show that objectives are not being met, an action plan needs to be established. Even when objectives are being met, areas for improvement can be noted. And as a project proceeds, the metrics used initially can be replaced with more sophisticated models that can measure progress more precisely and perhaps in a more valid or convincing ways. For this example, actual cost savings might be able to be measured in the future.

Publish Results and Collect Feedback

Have a plan for the metrics project. Consider

- Publishing the results of metrics efforts, perhaps in metrics summary Web pages.
- Collecting needs information before and during the effort.
- Collecting feedback on the understandability and usefulness of the metrics data itself.

Metrics Plan

The overall plan for a metrics effort might include proposals to

- Adopt a standard usability questionnaire and track data and progress.
- Set usability objectives and measure behavior in usability testing.
- Use checklists to track design problems and outcomes.
- Track usability issues and resolution.
- Introduce metrics to resolve subjective discussions about usability.
- Attempt to use usability data to prioritize design changes.
- Attempt to use return on investment to prioritize design changes.

- Attempt to prioritize design changes based on usability impacts on revenue and cost savings.
- Attempt quantitative competitive analysis.
- Collect metrics on the usability engineering process to manage and improve the quality of the design and evaluation process.
- Collect metrics (and success stories) to promote good usability engineering practices in the company and industry.

REFERENCES

(1999). *Assessing Web Site Usability from Server Log Files*. (White Paper No. 734-995-1010). Ann Arbor, MI: Tec-Ed, Inc.

Bailey, B. P., Konstan, J. A., & Carlis, J. V. (2001). Measuring the effects of interruptions on task performance, annoyance, and anxiety in the user interface. *INTERACT*.

(2001). *Best practices in collecting web user feedback*. Opinionlab.

Bias, R. G., & Mayhew, D. J. (1994). *Cost-justifying usability*. Chestnut Hill, MA: Academic.

Brooke, J. (1986). *SUS—A quick and dirty usability scale*. Reading, England: Redhatch Consulting.

Forrester Research. (2001). *Getting ROI from design*. Cambridge, MA: Randy Souza.

Lewis, J. R. (1995). IBM computer usability satisfaction questionnaires: Psychometric evaluation and instructions for use. *International Journal of Human–Computer Interaction, 1*, 57–78.

Mayhew, D. J. (1999). *The usability engineering lifecycle*. San Francisco: Kaufmann.

Molich, R., Laurel, B., Snyder, C., Quesenbery, W., & Wilson, C. (2001, April). *Ethics in HCI. Extended abstracts of the conference on human factors in computing systems*. New York, NY: Association for Computing Machinery.

Nielsen, J. (2001, January 21). Usability metrics [online]. Available: useit.com.

7

Learning About the User Experience on the Web With the Phone Usability Method

Julie Ratner

Molecular Inc.

INTRODUCTION

In this chapter an efficient discount method called phone usability for collecting user experience data is outlined in a step-by-step manner. The research, tips, and techniques explained in this chapter confirm that, even when executed over the phone, with the use of expert facilitation and planning by information architects or usability specialists, this methodology is a viable option for accessing valuable end user feedback.

Phone usability is a 16-hour method that focuses on preventative and diagnostic analysis of usability issues at one third the cost of traditional laboratory usability testing. The description in this chapter is the result of years of refinement of the phone usability method. The purpose of this chapter is to present the "what" and "how to" of the technique, as well as the advantages and disadvantages of this approach when, for example, only 2 days are scheduled for user feedback on a web development project.

What Is the Definition of Phone Usability?

Phone usability is a remote data collection method in a 30-minute phone interview while looking at a prototype on the Internet. The user answers questions and completes tasks using a "live" web site and/or a prototype on the Internet. While using the *thinking aloud protocol* (Dumas & Redish, 1999; Nielsen, 1993, 1994) the participants give feedback in real time on the design, content, and architecture to a usability specialist who facilitates the session, and takes notes to analyze later for a concise report. The entire process typically involves five phone usability sessions or until patterns in the user experience begin to emerge, and when recommendations can result in a report within a 16-hour time frame.

Human factors and usability experts concur that it is optimal to conduct traditional usability testing with a combination of user-centered design approaches (contextual inquiry, field studies, remote testing, user profiles, etc.; Krug, 2000; Nielsen, 1993; Dumas & Redish, 1999) whenever possible. However, when budget or time constraints eliminate the possibility of employing statistically valid research design methods, phone usability is another option to consider. Results from phone usability sessions at Molecular, Inc., an Internet professional services company since 1994, have prevented expensive user-centered errors on many Web development projects, including retail, intranet, and financial Web sites. Monk (2001) rallied, "There is a need for more techniques that fit Nielsen's requirements for discount techniques. These are lightweight procedures that can be learned in a day or so and take only man-days to apply." After researching discount usability methodology, this chapter documents the phone usability method to share the value of this approach for experts wishing to implement or coach other specialists (with backgrounds in technical writing, engineering, interface design, or marketing) to use this technique.

HISTORICAL OVERVIEW

Some terms connoting 'discount usability' and 'guerilla warfare testing techniques' as quick fixes to satisfy the most skeptical development teams have become popular in user interface design circles. Terms such as 'hallway methodology' (Nielsen, 1997), and slogans such as "Testing one user is 100% better than testing none" (Krug, 2000), confirm the assumption that many usability professionals face similar challenges of time limits and budget cuts, particularly in the domain of Web development. This phone usability methodology attempts to answer the question: What is the least amount of staff time and cost necessary to deliver usability results of value for developers and designers on the team?

How Does This Method Compare With Other Discount Usability Techniques?

In 1994, Nielsen published a paper on "discount usability engineering" for development teams that need a quick, inexpensive, valid, and reliable usability test based on four techniques: user and task observation, scenarios, simplified thinking aloud, and heuristic evaluation. In that same year, Bias and Mayhew (Nielsen, 1994), coined the term "Guerrilla HCI" in a chapter subtitled "Using Discount Usability Engineering to Penetrate the Intimidation Barrier." This chapter dispelled the myth that involving users is synonymous with cost overruns and timely project delays. The researchers challenged the preconception that usability was expensive and time-consuming, and that testing needed to be conducted by experts with Ph.D.s. To quote Nielsen (1994), "(C)omputer scientists are indeed able to apply the thinking aloud method effectively to evaluate user interfaces with a minimum of training, and that even fairly methodologically primitive experiments will succeed in finding many usability problems." Since the early 1990s, many more researchers (Krug, 2000; Nielsen, 1989, 1999; Ratner, 1999, 2000) have preached the benefits of discount usability testing methods, including hallway testing by co-workers.

Who Can Conduct Phone Usability?

Bias and Mayhew (Nielsen, 1994) wrote, "So we can conclude that running even a small, cheap empirical study can help non-human factors people significantly in their evaluation of user interfaces." It is critical to both the Web development industry and to usability practitioners that volunteers feel as if their experiences are valuable to the Web development teams and that their feedback, both positive and negative, is collected methodically and analyzed carefully. Ideally, phone usability, like all other usability test methods, will be conducted by a trained usability specialist.

An expert with finely honed analytic and listening skills is invaluable to a development team under a strict timeline to launch a Web site.

> "Many teachers of usability engineering have described the almost religious effect it seems to have the first time students try running a user test and see with their own eyes the difficulties perfectly normal people can have using supposedly "easy" software. Unfortunately, organizations are more difficult to convert, so they mostly have to be conquered from within by the use of guerrilla methods like discount usability engineering that gradually show more and more people that usability methods work and improve products." (Nielsen, 1994)

Phone usability, like many guerilla methods, can be employed at any organizational stage: skepticism (stages 1–2), curiosity (stages 3–4), acceptance

TABLE 7.1
Comparison of Phone Usability and Phone Interviewing

Phone Usability	Phone Interviewing
Pros	
• Able to approximate load	• Hardware connection not necessary
• Task-focused questioning	• Open-ended questioning
• Prototype focuses dialogue on specific usability issues	• Dialogue focuses on past experience; typically no visuals unless viewing a facsimile of prototype
• Contextual pre- and post-questions based on tasks	• General contextual inquiry
• Employ thinking-aloud technique to follow path	• Requires vivid memory of path
• Listen to reaction to current and prototype design online	• Rarely collect feedback on any prototype; occasionally listen to reaction to monochrome facsimile
Cons	
• Conduct qualitative and quantitative inquiry	• Conduct qualitative inquiry
• Access only to verbal feedback, lose expression of facial and body language	• Same
• Limited to 30-minute session to restrict number of tasks	• Limited to inquiry questions
• Unable to record cursor movements	• Same
• May lose observer involvement, because only audio	• Same
• May not capture exact quotes because usually no audio tape	• Same
• Limited time in 30-minute session for follow-up comments	• Same
• Scheduling is inflexible because of 16-hour timeline	• Same

(stages 5–6), or partnership (stages 7–8, Ehrlich & Rohn, 1994), to ease the intimidation barrier and facilitate the transition of development and management teams toward awareness. There may be expensive user-centered issues based on assumptions about the end users that are incorrect. These assumptions may be confirmed or unconfirmed in only one round of phone usability testing. As listed in Table 7.1, phone usability involves task-focused questioning and so the data is more convincing than using phone interviewing alone.

What Is the Cost Difference Between One Round of Phone Usability and Traditional Usability Testing?

Whenever the issue of cost surfaces in digital strategy meetings, inevitably the value of discount method is debated. In Table 7.2, a high level cost comparison highlights the 3 to 1 expense ratio per iteration of phone usability as compared to traditional usability lab testing.

One expense not included in Table 7.2 is the cost of equipment maintenance, lab equipment (scan converters, video cameras, video cassette recorders, and digital

TABLE 7.2
Cost Comparison of Phone Usability and Traditional Usability Lab Testing

Expense	Phone Usability	Traditional Usability
Recruiting	Typically by e-mail, phone, or poster	Outsource for $125 per recruit × 5 = $625
Incentives	$10 gift certificate × 5 = $50	$150 × 5 = $750
Consent Forms	Not required	Required
Users	5 users × 30 min = 150 min	5 users × 90 min = 450 min
Preparation	2 tasks or scenarios	5 to 7 tasks or scenarios
Prototype	2 hours × $200 = $400	5 hours × $200 = $1,000
Staffing	Facilitator × 4.5 hrs × $200 = $900	Facilitator and note taker × 15 hr × $200 = $3,000
Observers	Recommended	Recommended
Data collection	Notes	Notes and video tape
Debriefing	1 hour × $200 = $200	2 hours × $200 = $400
Report	3 hours × $200 = $600	5 hours × $200 = $1,000
Total time	Two-day minimum + $2,150	One week minimum + $6,775

mixers), if used regularly, requires a budget for maintenance. By contrast, phone usability requires only a phone and headset or speakerphone equipment with minimal additional servicing costs. There is one primary negative trade-off to phone usability testing that may not be immediately apparent when looking at the above cost comparison: no face-to-face observation of users. When the development team is totally dependent on participant's ability to use the thinking aloud protocol and there is one fourth time for tasks, there is a high risk of decreased involvement and less trust in the results. In each of the steps described below there are advantages and disadvantages that will be important to evaluate against each project business goal to determine whether this method is appropriate to use. Following is a complete list of phone usability steps:

1. Write a usability plan, a test plan, rehearse the test plan, and then modify questions and tasks as needed;
2. Create and deliver prototype online;*
3. Recruit users via phone, e-mail, or posters;
4. Send e-mail confirmation to scheduled users;
5. Prepare phone usability observers. Find location for testing with a POLYCOM Sound Station Premier™ 'spider' phone and one computer.**

---

*Step considered optional.
**Also add a slave monitor, if situated in a large conference room with more than three observers. Copy a stack of test plans along with the handout on "How to be an Observer" and debriefing form and provide clipboards, pens, and post-its to encourage note-taking and active participation among observers.

6. Conduct a minimum of five interviews;
7. Collect post-observation form sheets and debrief with the team. It is optimal to schedule the debriefing immediately following interviews, whenever possible;*
8. Analyze data, write and deliver a summary report;
9. Present findings to key stakeholders. Document questions and answers from stakeholders and e-mail decisions to the team.*

ADVANTAGES AND DISADVANTAGES OF PHONE USABILITY

Whenever usability methods are scrutinized, it is important to delineate both the advantages and disadvantages. As Nielsen (1994), Monk (2001), and Krug (2000) have concluded, if the choice is whether or not to involve users, it is preferable to employ whatever discount usability techniques are available to improve the user experience, given time and budget constraints. User experience evaluation, also called an heuristic evaluation, or a "crit," is valuable to measure the architecture and design of a site against best practices. As the Comparative Usability Evaluation (CUE) research studies have proven (Molich, 2001; Molich et al., 1999), evaluations are best conducted in *combination* with other usability methods with diverse end users because they "Provide opportunity to collect various types of data and provide more flexibility in terms of what you can do with the data" (Molich, 2001).

The phone usability method is based on simplified traditional discount usability techniques published by Nielsen (1993) and Dumas and Redish (1999). The purpose of listing the steps is to increase awareness of the benefits of conducting usability over the phone. Examples of *portions* of the recruiting script, e-mail invitation, confirmation, test plan, and test results summary are included from a phone usability study of the Molecular, Inc. Web site to help speed the adoption of this technique by corporate and academic practitioners.

The phone usability method, like many practical user experience techniques, relies on common sense and analytic and interviewing expertise along with basic behavioral psychology techniques for 'drawing out' non-verbal participants. General practical testing tips from Krug's (2000) book, "Don't Make Me Think!" directly apply for phone usability as well:

- If you want a great site, you have to test;
- Testing one user is 100% better than testing none;
- Testing one user early in the project is better than testing 50 near the end;
- The importance of recruiting representative users is overrated;
- The point of testing is not to prove or disprove something. It's to inform your judgment;

TABLE 7.3

Approximate Hours Per Step in Phone Usability Method

Step	Method	Approximate Hours
1	Preparation	4.0 hours
2	Recruiting users	2.0 hours
3	Phone usability interviews	4.5 hours
4	Analysis	2.5 hours
5	Summary report	3.0 hours
	Total	16.0 hours

- Testing is an iterative process;
- Nothing beats a live audience reaction.

There are tradeoffs among at least three variables—time, quality, and cost—in each step of phone usability technique. There are repeated decisions, such as expanding time, cutting back on the number of users, increasing the level of involvement from developers and designers, and decreasing the design time of a prototype. Each decision impacts the pace of the phone usability session and the quality of results as well as the return on investment (ROI) for the project. In the schedule outlined in Table 7.3, emphasis is on preparation and data collection to fit an average 16-hour, one-person day time line.

'Speed of Light' Web Development Cycles

A speed of light Web development cycle typically refers to a project when one or two days to collect end user feedback are allotted, and may mean the difference between high return on investment (ROI) with immediate positive impact versus zero or negative ROI for a development team.

The time required for completing all the steps in the phone usability process is variable. It is often a difficult task for an information architect or a usability specialist to determine how to divide up 16 hours, whether preparing a test plan, recruiting users, conducting five phone usability interviews, analyzing data, or writing a summary report. It depends on a number of complex variables whether a 16-hour estimate is realistic. Sixteen hours is the *minimum* average, depending on these key variables:

- Preparation time of the prototype
- Familiarity with the Web site
- Access to users (during the day)
- Skill of team at analysis and summarization of results
- Assistance by the team for recruiting
- Availability of observers for note-taking and debriefing of the data

When Is the Phone Usability Method Valuable?

The feasibility of completing a phone usability event in 16 hours is determined by matching the business and usability goals and evaluating the trade-off of adding or omitting tasks and questions for the users in a 30-minute test plan. An inverse relationship exists between time and length of report and value and impact to both the project development team and business stakeholders (management or clients). For example, the longer the development team or stakeholder waits for a usability findings report, the less impact and lower perceived value of its content. This provides a general explanation why the following 16-hour schedule allows for four hours of preparation and only three hours for writing the summary report. In my experience, a summary report delivered immediately following phone usability interviews (preferably the same day) is typically read thoroughly as compared with a comprehensive report delivered within the week, which may only be skimmed.

Phone Usability as Compared to Other Discount Usability Techniques

Nielsen published a table in 1993 summarizing the methods of the engineering life cycle and their pros and cons. Even though three discount methods are outlined in this summary, it is important to reiterate that *testing does not rely on a single method to the exclusion of others.* Phone usability is added to Nielsen's list (1993) in Table 7.4 to a condensed version of many techniques available to a usability specialist when five or fewer users are required.

"Since the early 1980s, think-aloud protocols have been used very successfully in understanding users' problems" (Dumas & Redish, 1999). Interviewing typically involves open dialogue and observation of end users is often referred to as contextual inquiry. Phone interviewing is frequently designed to be a "user-lead

TABLE 7.4
Phone Usability Method Added to Compare With Advantages and Disadvantages
of Two Discount Usability Methods from Nielsen (1993)

Method Name	Users Needed	Advantages	Disadvantages
Thinking aloud	3–5	Pinpoints user misconceptions; cheap test.	Unnatural for users; hard for expert users to verbalize.
Phone interview	5	Flexible, in-depth attitude and experience probing.	Time-consuming; hard to analyze and compare.
Phone usability	5	Fast, inexpensive for targeted user feedback—on a prototype and/or current site.	Limited to 2–3 issues. Phone "observation" requires users to think aloud.

TABLE 7.5

Advantages of Phone Usability

1. Simplifies the recruiting of users when sessions are conveniently on the phone for 30 minutes.
2. Contextual pre- and post-questions confirm or disprove assumptions about audience.
3. Task-focused questioning helps focus user on the most severe usability issues.
4. Able to "watch" path as user completes task from thinking aloud protocol.
5. Prototype focuses interview questions on—solutions being considered.
6. Collects immediate user reaction to prototype design.
7. Documents load time on personal hardware and user reaction to the Web site on user's own system.
8. Real user feedback with no travel time required.
9. Training opportunity for developers and designers to interact with users in a scripted and structured session.
10. Limited amount of data is collected in 30 minutes, so less time is required for report summary.
11. Quality assurance testing of user's hardware with various browsers and connection types.

TABLE 7.6

Disadvantages of Phone Usability

1. No face-to-face contact or video means lose facial expression and reliant on thinking aloud protocol.
2. Restricts number of usability issues to tackle in a 30-minute session.
3. Depends on thinking aloud protocol for path and specific user experience information because there is no remote automated recording of cursor movements.
4. Less active involvement and participation of observers, because there is only audio connection with end user, no video in testing space.
5. Dependent on note-takers jotting down paths and comments on sticky notes or on debriefing form; rarely record on audio to later transcribe.
6. Tight scheduling with no room for cancellations in a 16-hour schedule means frequent recruiting of users who do not meet the recruitment criteria.
7. Collects mostly qualitative data and very little quantitative data.
8. Get signatures on non-disclosures infrequently because most prototypes do not contain proprietary information and not video or audio taping to require consent forms.

conversation" (Dumas & Redish, 1999). Phone usability combines thinking aloud protocol, phone interview skills, and benefits from some of the "real time" advantages of observation of remote testing techniques (see Tables 7.5 and 7.6). Unlike the automated testing of remote testing services, which rely on users typing explicit comments after each task, the phone usability method incorporates the thinking aloud protocol and allows a usability specialist the chance to request clarification directly with the participant:

> Having users think out loud while performing any task, from reading a text to working with a product, is also having the user *give a verbal or think-aloud protocol* . . . The best talkers sound as if they are giving you a stream of consciousness and they also add their interpretation of events. The worst talkers say almost nothing or, worse, mumble. Most participants fall somewhere in between. (Dumas & Redish, 1999)

The risk as noted by Dumas and Redish (1999) is that "(t)est participants vary in their ability to tell you what they are thinking while they work."

TIP: A good strategy is to give participants think-aloud instructions and one or two warm up exercises to model the process.

PHONE USABILITY METHOD
STEP-BY-STEP

Step 1: Write Usability Plan and Test Plan

Preparation and planning are essential to the success of any usability effort. The plans are especially critical to successful 16-hour phone usability testing events. The fundamental goal of the usability plan is to identify the business goals, assumptions, development, and design questions to formulate tasks and pre- and post-questions in the test plan. This in turn helps to outline a digital strategy to maximize the value and ROI of the user experience data to meet the top business priorities.

The following example is a portion of a phone usability test plan on the use of Flash animation on global navigation appearing on the Molecular.com Web site to illustrate how pre- and post-task questions provide context for the user behavior observed during the task. During the session, the motivations help remind the team of the focus of the user experience issues specific to the task. The more precise the task in terms of motivation, the less likely the team will stray from the test script onto tangents unrelated to the immediate business goals.

Sample From a Section of a Phone Usability Test Plan

Pre-task question: What was your first impression of the Molecular site?
[Motivation: To find out what users recall about the design and content of the site: What adjectives do they use to describe the Web site if they visited it since the brand transfer in May 2001? Did they mention dynamic Flash or static HTML? What was their perception of Molecular's use of Flash technology throughout the site, if any? Do they mention the "cool" dynamic technology on the site?]

Task: Show me three different ways to get to a case study from the current Molecular Web site.
[Motivation: To find out which is the dominant path for finding case studies; Does the nomenclature and architecture make sense under "clients" or do users look under "innovation" or "expertise?" Do they have Flash plug-ins loaded? How do they get back to home, using the logo or the 'home' link on the utility bar or the browser's 'back' button? What terms do they type in the search field? What do they think of the search results?]

Task with prototype: [Mouse over company, client, and expertise on the global navigation on the site.]
What do you think of the speed and the feedback on the bar?

Post-task question: Once you figure out how the navigation works, it is usable?
[Motivation: To find out what users think of the design and timing of the current and prototype navigation.]

TIP: Get onto the schedule on an iterative basis, as frequently as the development team and project manager permit—on a weekly, bimonthly, monthly, or quarterly basis. It is better to be positioned on the schedule frequently and later reformat a round of traditional lab testing into phone usability testing if the schedule is delayed or suddenly shortened.

Step 2: Creation and Delivery of the Prototype

Most phone usability participants appreciate a sneak preview and feel privileged to see a prototype as well as give their opinion. Occasionally, no prototype is tested and only the current site is reviewed. Although this is not problematic, in a way it is a missed opportunity to get feedback on innovative ideas that could become more clearly defined with user input.

Prototypes (see Fig. 7.1) for phone usability sessions can be created as HTML, Microsoft Excel, or PowerPoint files (as JPEG static images or PPT slides), so that they look purposely distinct from "live" HTML pages. The more like "sketches" and design "sneak previews" they appear, the more feedback users are likely to share them. Because wire frames and full-blown design comps usually seem polished and "complete," users are less likely to believe the design is a work in progress.

Typically, only one part of the prototype is shown in detail (see Fig. 7.2) and the rest is merely blocked out. This helps both the facilitator and the participant focus on the targeted usability issues without dialoguing on tangential issues.

There are various options for delivering a prototype for a phone usability session, including faxing, e-mailing a Microsoft Excel or PowerPoint attachment, or converting Microsoft PowerPoint into HTML slides and hosting them on a temporary area of an Internet. The temporary area needs to be accessible, beyond any firewall and only "live" for the 16-hour duration. Then it can be removed. The prototype page should not be linked to any other part of the Web site and is usually in a separate directory called "test" or "usability."

It is rare that active links are required on phone usability prototypes. Because of time constraints during the preparation phase, it may actually be a detriment

FIG. 7.1. Complete detailed prototype to collect feedback on the entire design of the Molecular.com home page.

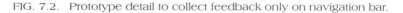

FIG. 7.2. Prototype detail to collect feedback only on navigation bar.

company
clients
expertise
careers
innovation
news
history
management
values
community

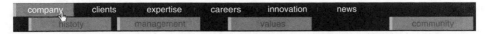

FIG. 7.3. Original Flash animation of sub-navigation buttons "flying in" right to left under "company."

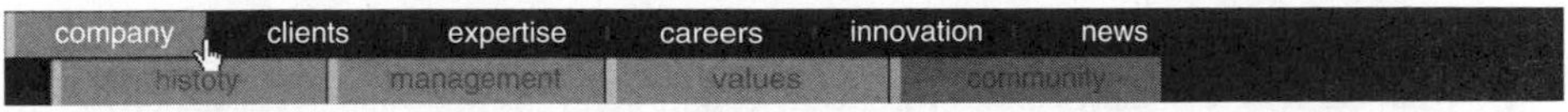

company
clients
expertise
careers
innovation
news
history
management
values
community

FIG. 7.4. Revised Flash animation of sub-navigation buttons fading in under "company."

to the participant if the links work. In the rare event that functionality is critical for the feedback, XLS and PPT files can both include hyperlinks. By contrast, when animation is being tested, working Flash files can be placed on Web pages as links to test animation, speed, directionality, mouse response, and interaction design feedback (see Figs. 7.3 and 7.4).

Paper mock-ups scanned and hosted on the Web have not been highly successful in comparison with PPT, XLS, and HTML prototypes because of the lack of legibility. Although graphic designers may correct this issue, the time involved may not be warranted relative to the total duration of a round of phone usability testing.

TIPS:

1. Collaborate with graphic designers on the preparation of phone usability prototypes to determine what level of detail in the design is necessary. Whenever there is doubt, refer to the task and its motivation to focus on the detail critical for comprehension of the user experience. Be forewarned: Prototypes that are too detailed tend to waste time and to collect extraneous data that is not "mission critical" to the development team (see Figs. 7.1 and 7.2).

2. It may make sense to combine two delivery methods in case one mode fails. For example, if a prototype file is too large for a participant's e-mail folder, revert to discussing a facsimile. If a volunteer is at a vacation home and the available computer does not have Microsoft PowerPoint, convert the PPT slides to JPEG images for viewing each slide from any Web browser.

Step 3: Recruit Users via Phone, E-mail, or Posters

A recruiting script helps to find the "best" participants in a tight 16-hour time frame. In phone usability the key is to make sure you have the best users possible given the recruit time available. The description in the recruiting script identifies the target user to best *match* the usability goals of the test plan. The script starts as a list of qualifying and disqualifying criteria and may later expand into a user profile later in the overall project. Two examples for recruiting for phone usability are listed below to help clarify the importance of selection criteria:

> **Example 1:** Recruit five women between the ages of 25 and 45 who have purchased from the XYZ company online or from any candle company once in the past year, if ever.
>
> **Example 2:** Recruit clients and prospective clients of Molecular who are business executives in decision-making positions, that is managers, vice presidents, CIOs, CEOs, CFOs, and so forth.

Depending on the type of Web site you are testing, recruiting may be as easy as setting up a notice in the lunchroom, on the bulletin board, in a cafeteria in a local restaurant or office building, or sending an e-mail to your colleagues, family, and friends.

Krug (2000) published a list of reminders relating to recruiting that could also be applied to phone usability recruiting:

• Recruit loosely and grade on a curve. [Author's note: To save precious time, any team member can recruit as long as the recruiting script is used to set realistic and consistent expectations among the volunteers.]

• It's usually not a good idea to design a site so that only your target audience can use it

• Experts are rarely insulted by something that is clear enough for beginners

• Offer a reasonable incentive. [Author's note: In phone usability there is a two-hour maximum for calls, e-mails, faxes, or posters, so there needs to be a direct relationship between the stringency of the recruiting criteria and the desirability of the incentive. For example, when I needed to recruit candle buyers for phone usability, an incentive was unnecessary because the criteria was general; nothing was needed to motivate these volunteers.]

• Keep the invitation simple

• Avoid discussing the site beforehand

• Don't be embarrassed to ask friends and neighbors. [Author's note: I only recruit Molecular company employees as usability volunteers and schedule them as "standby" if we can't recruit more typical end users, because Molecular staff are too savvy from work experience in the Web development industry.]

This is an example of the Molecular phone usability e-mail invitation sent to clients and prospective clients from a database from our business development group to help recruit phone usability volunteers to give feedback on Molecular.com.

TIPS:

1. A combination of e-mail and phone recruiting works best. As a rule, direct calls are the fastest way to recruit; however, depending on the Web site being tested and the recruiting criteria, e-mail invitations can quickly supplement the list of names of volunteers, particularly if calls need to be made on Friday afternoon, typically a challenging time to recruit.

2. Outline in a telephone script or e-mail invitation explaining exactly why you are recruiting volunteers and what will happen to the feedback you collect, as well as the incentives for participation.

3. Use appropriate language for your volunteers. Reiterate that it is the interface being tested, *not* the volunteer. Although I prefer to use the word phone usability session in place of the word test and many times replace "keywords" with 'Web sites,' I like Parkins' (1998) invitation script: *"The 'test' is actually fairly short and is carried out on a very informal basis. And remember, it is YOU who are evaluating the software, NOT the other way round, so don't let the word 'test' mislead you! If you are interested in any of the following keywords: Design Patterns, Object-Oriented Concepts, VRML, Usability Engineering, Human-Computer Interaction, and User Interface Design, then this should be of interest to you!"*

4. Professional recruiters recommend that more volunteers be scheduled than the minimum required because some will always cancel. For that reason, in the confirmation e-mail it is important to be specific how to re-schedule or cancel the session, so that the development team can be alerted whenever possible.

5. In the e-mail invitation, include a sentence on how to prevent other e-mail invitations for other usability events to avoid annoying volunteers.

6. Provide a recruiting script if multiple people are calling for volunteers so that all participants have the exact same expectation as to the purpose of the phone usability session.

7. Provide incentives, even if only a $10 e-mail gift certificate, free lunch, a corporate polo shirt, toy, or mug to help reinforce the fact that user feedback is a valuable part of the Web development process at your company.

Step 4: Send an E-mail Confirmation to Scheduled Volunteers

The confirmation e-mail helps prevent last minute cancellations or "inappropriate" recruits from participating. In this e-mail, the reasons why certain volunteers have been selected are listed and why the company is conducting phone usability is repeated (in general terms). That way, if an individual didn't clearly understand the purpose of the session or accidentally confused the purpose with a marketing survey or business development call during recruitment, he or she can easily cancel.

The e-mail contains the time and date of the phone usability interview along with a description of each step of the session. Included in the e-mail is text to reassure the volunteer that it is the Web site and the prototype that are being evaluated. A consistent tone in this message is critical during recruitment and in the confirmation e-mail; it is key to reiterate that it is the Web site, not the user's knowledge of the Internet or the company Web site in question, that is being tested.

This needs to be repeated for users many times throughout the phone usability interview, starting in the confirmation e-mail. The text in the e-mail also explains the thinking aloud protocol using an analogy of a radio announcer. Volunteers of virtually any age, from 13 to 90, can understand the concept of the verbal play-by-play of a baseball game.

TIPS:
1. Confirm the date and time and provide specific details in a follow-up e-mail and embed Web links or prototype attachments.
2. Use an analogy relevant to the age of your volunteers to explain the thinking aloud protocol.

Step 5: Preparing Phone Usability Volunteers for Observer Participation

It is recommended that the team members read about and experience interviewing and note-taking through role-playing the best practices during rehearsals of the test plan. That way the team is confident during the session and enjoy asking open-ended questions at its conclusion. Best practices for observers are quite similar to standard lab testing rules of conduct with a few exceptions, because the interaction takes place over the phone. It is helpful to prepare users for any background noises (for example, laughter from an adjacent cafeteria) or to explain what distractions may be heard (for example, jackhammer sounds for building repair or a beep if a developer from the New York office joins the conference call). Promise to announce team members if any enter the room and indicate that they may ask questions at the end of the session if there is time.

The key is that observers are familiar with the best practices so that each session is facilitated using the same script in a consistent neutral tone. When observers are co-authors of the test plan and participate in the rehearsal of the questions and tasks, and are clear about the motivations for each task, there are rarely interruptions during a phone usability session. Occasionally, a volunteer or an observer interrupts the flow of the session and the team diverges from the key tasks. When this happens, the facilitator needs to re-focus the session toward the usability goals. Because there will be many opportunities to discuss with the observers after the session, the top priority is to reassure the user when the session is redirected onto the topic.

Portion of an Observer Best Practices Handout

This is one of a series of test sessions—all observers will share ideas and their impressions at a debriefing to build a "to do" list and make the next Web interface design more intuitive:

During the test, the observer should notice:
- Which interface clues did the user understand? Which did they misunderstand?
- Which parts of the Web interface design were never noticed?
- Was the user aware when he or she made a mistake? Did the user recover from error easily?

Observe:
- Maintain **neutral** phone etiquette (no sighs, tapping pen, shifting during note taking, passing notes).
- Speak to the individual the same way when a mistake is made as when something is correctly done; when a mistake is made, continue as if it hadn't occurred.
- Never make participants feel stupid.
- Don't assist them when they struggle (we learn more when they are lost and see how easily they recover).

Help during testing:
- Limit your prompts to encourage them to think aloud.
- Provide hints only when the customer asks for them. (Don't give them the path to find information, answer a questions with a question when you can, to elicit more information about what they are looking for.)

Discussion:
- Do an "internal body scan" before asking a question to register whether you are reacting to the user's experience or using *neutral language*.

TIPS:

1. During the phone usability session, the observer should listen to the user, take notes, and pass questions to the facilitator if unsure when a question is "leading".
2. If there is time at the end of a phone usability session, encourage an observer to try out a *new* question *not* on the test plan.
3. Practice how to phrase open-ended questions and explain how much more valuable descriptive answers are in the context of a task beyond yes or no answers.

Step 6: Conduct Phone Usability Interviews, Document Feedback Debrief

The first priority when conducting phone usability is to ensure that the volunteer enjoys the 30-minute session and is willing to participate in the future—at the participant's convenience. The second priority is for the development team to collect the usability data it needs to solve the business problem and to confirm during the debriefing that the solution will continue to maintain or improve the user experience on the Web site. The third, and final, priority is to educate, mentor, and teach the end user, the stakeholder, and development team that interactions with end users, even during a discount phone usability session, will help increase ROI and improve the user experience.

Phone usability testing requires flexibility and resourcefulness because an equal number of unexpected distractions can occur over the phone, as in a laboratory. For example, during one session, an engineer was thinking aloud and completing the tasks on his laptop while watching his 3-year-old son. The volunteer started laughing and interrupted the session because his son was playing, fully dressed on his portable potty, and got stuck. Fortunately, no one was hurt and the session continued. Like all usability events, starting and ending on time, being prepared and respectful when interruptions occur, and using diplomacy when the unexpected does happen is crucial.

Sometimes minutes before a session the volunteer's system crashes, the network goes down, an emergency meeting is scheduled, or personal child care responsibilities take precedence. The most successful way to ensure repeat and ongoing participation among volunteer usability testers over the phone is to reassure them of future opportunities to get involved when they cancel.

The debriefing meeting empowers the development team in decision-making, when the test plan or prototype needs to be adjusted during the first round of phone usability. Typically, only the facilitator attends all the phone usability sessions and most observers only participate in one of the five. For that reason, it may be challenging for the development team to predict the patterns in the data and make decisions without a team debriefing. A productive debriefing session is easy to achieve with a clear and concise agenda.

As with the preparation for an interactive design review meeting (Ratner & Gondoly, 1998), identify the issues, set a time limit for discussion for each issue, identify possible solutions, and add new solutions as they surface. Avoid repeating unnecessary background in the agenda. Instead, focus on the part of the test plan or prototype not yet achieving results and determine with the development team whether there are other solutions that would benefit from user feedback. Include screen captures and captions to assist team members less familiar with the user interface (see Figs 7.3. and 7.4) and to document where users had problems.

Sample of One Portion of Data Collection Log Format

Pre-task question Do you remember what your first impression of the site was?	*Motivation* To find out how participants describe the design and content of the site.	*User #2* "When launched new brand [noticed] impact of manifestation red—black—fast moving Flash—dynamic home. That was a negative reaction—too fast—dizzy."	
Task on current Web site Find out more about the non-profit social mission Web projects that Molecular has completed.	*Motivation* To find out whether navigation is difficult on the community tab (because it's the farthest right), if community makes sense participant's reaction to the social mission work, and if they use the search; what terms they search on.	*Path* 1 > home (utility bar) 2 > news 3 > company 4 > site map 5 > company 6 > community Total of 4 tries	*Quotes* 3 > "This is what drives me nuts, rebuilding that's [the] dizzy part." "I have had no caffeine. Not in a hurry, could take me more tries. Would try site map next." 4 > "I'm a site map person."
Task on prototype Think aloud to compare navigation on both Flash bars to look for info on case studies.	*Motivation* To find out what participants think of the speed and mouse manipulation, whether they notice the bar squares, and their experience in refreshing the cycle		"Notice new mouse over clients 2nd level coming in from right less distracting, more logical if builds left to right."
Post-task question What do you think of the speed and the feedback on the bar?		"A much better—faster—feeling of being faster—tighter—not as distracting because filling in rather than flying in (don't see gaps in 2nd level)."	

Debriefing serves dual purposes: it helps expand awareness about discount usability within the organization, and it involves and empowers the development team in the phone usability effort so that there are fewer surprises when the results are placed on the project Extranet.

TIPS:

1. The facilitator of the phone usability session is responsible for reassuring the volunteer and paraphrasing the user's comments to ensure the feedback was properly interpreted.

2. Remind them that *only* those who attended all the interviews have a big picture of the user experience. Explain to observers that they need to share ideas at the debriefing to build a "to do" list in order to make the next Web interface design more intuitive.

3. Use Dray's (2001) post-observation debriefing sheet prior to group discussion. That way, even if developers and designers are unable to attend the debriefing meeting you have their immediate feedback and get a glimpse of what they learned and what in the test plan needs to be edited prior to the next interview.

4. The debriefing form provides immediate feedback to the facilitator regarding what the observers noticed, whether the 30-minute interview expanded awareness about the user experience, and whether the test plan needs to be refined prior to the next phone usability session.

Step 7: Analysis and the Summary Report

The analysis and summary of findings after one round of phone usability testing has standard sections, such as executive summary, user quotes (positive, neutral, and suggestions for improvement), summative statistics: issues ranked by severity, frequency, and persistence, and recommendations. Sometimes there are unexpected comments relating to the phone usability methodology. These are typically highlighted in an additional section or in an appendix.

Think carefully about the audience before delivering a comprehensive phone usability report (Dumas & Redish, 1999). Often there are multiple audiences that require different levels of detail and who measure return on investment (ROI) using different metrics. Plan ahead when interviewing, taking notes, and debriefing with observers if there are particular qualitative and quantitative metrics that will confirm ROI for your stakeholders and then highlight them in the summary report (Buchanan & Lukaszewski, 1997).

Portion of Summary Report

Executive Summary

Round three of phone usability of Molecular.com compared the current navigation on the site with a revised Flash navigation prototype with six clients. Flash navigation was strategically targeted to resolve usability problems relating to:

- Speed and directionality of Flash navigation buttons
- Perception of the application of Flash technology on global navigation bar
- Ease of navigation using mouse on global navigation bar

100% are current clients using Molecular services, and 33% (or three of the six participants) had visited the site since May when the brand changed from Tvisions to Molecular, Inc.

Phone Usability Results

- 100% found the Flash navigation prototype **usable**. 84% perceived the prototype to be faster for finding information; the other 16% were neutral about the change. User quote: "The Flash is less intrusive [on the prototype]; movement is subdued and not quite as frantic. The individual buttons slide in together, as a unit. Seems to me you are improving it."
- 100% were able to move the mouse to the sub navigation options on the first try as opposed to 66% on the current site. One client stated, "It took two tries to get to the button on the current site. I missed the moving target. One bar wasn't moving the way the other bar does. The redesign is very stable, easier to navigate."
- Flash navigation prototype is **easier-to-use**. 84% of clients reacted positively to the movement of the sub-navigation bars. One client said, "The timing is right in the prototype; nice and smooth movement. I like how they fade in."

TIPS:
1. Think ahead about whether certain data entry forms or report formats can help speed analysis and visualization of the data for the team to summarize.
2. Whether or not time and budget permit the delivery of both a summary report and a comprehensive report, plan to be on-hand to answer specific questions and document any stakeholder questions as well as your expert answers for the entire team to read after the meeting.

CONCLUSION

"A basic problem is that with a few exceptions, published descriptions of usability work normally describe cases where considerable extra efforts were expended on deriving publication-quality results, even though most development needs can be met in much simpler ways." (Bias & Mayhew, 1994). Have you ever heard a developer confirm, "I know there are usability problems with this Web site, but we don't have two weeks in the schedule to talk to users?" When a time constraint surfaces in your next development meeting, consider using the two-day phone usability technique.

144   RATNER

ACKNOWLEDGMENTS

Special thanks go to my co-workers at Molecular, Inc., including Megan Nickerson and Laverne Phillips. I send my heartfelt gratitude to my volunteer editors for the entire book: Michael and Lynn Thornton, Deborah Fillman, Peter Denisevich, Anne Fassett, Grace Induni, Ruth Stillman, David Forrester, Maura Nagle, Heidi Ratner-Connolly, and Mark Gens. I also send appreciation to my acupuncturist Min Zhu for helping me to stay healthy and to my family and friends for their patience during the summer and fall of 2001. My sources of inspiration were my toddler friends, Emily and Isabel. I encourage readers to share improvements-to this method with me at jratner@onebox.com.

REFERENCES

Buchanan, R. W., & Lukaszewski, C. (1997). *Measuring the impact of your Web site*. New York: Wiley.
Dray, S. (2001). *Understanding users' work in context: Practical observation skills*. Unpublished tutorial notes from a CHI2001 conference, Seattle, WA.
Dumas, J., & Redish, J. (1999). *A practical guide to usability testing* (Rev. ed.). Portland, OR: Intellectbooks.
Ehrlich, K., & Rohn, J. (1994). Cost-justification of usability engineering: A vendor's perspective. In R. G. Bias & D. J. Mayhew (Eds.), *Cost justifying usability* (pp. 73–108). Boston: Academic.
Krug, S. (2000). *Don't make me think! A common sense approach to Web usability*. Indianapolis, IN: New Riders.
Molich, R. (2001). "Advanced usability testing methodology," *UIE6 conference proceedings*. Cambridge, MA: User Interface 6 East Conference.
Molich, R., Kaasgaard, K., Karyukina, B., Schmidt, L., Ede, M., van Oel, W., & Arcuri, M. (1999). *Comparative evaluation of usability tests* (CHI99 Extended Abstracts, pp. 83–84). Denver: ACM.
Monk, A. (2001, August 4). *Lightweight technologies to encourage innovative user interface design* [online]. *Systemconcepts*. Available: http://www.system-concepts/masterclass/andrewpaper.html
Nielsen, J. (1989). Usability engineering at a discount. In G. Salvendy & M. J. Smith (Eds.), *Designing and using human-computer interfaces and knowledge based systems* (394–401). Elsevier Sciences Publishers, Amsterdam.
Nielsen, J. (1993). *Usability engineering*. Boston: Academic.
Nielsen, J. (1994). Guerrilla HCI: Using discount usability engineering to penetrate the intimidation barrier. In R. G. Bias & D. J. Mayhew (Eds.), *Cost-justifying usability* (pp. 245–270). Boston: Academic.
Nielsen, J. (1997). Discount Usability for the Web http://www.useit.com/papers/web_discount_usability.html
Parkins, J. (1998). *A detailed user evaluation trial of a 3-D learning environment for object-oriented design patterns* [online]. De Montfort University. Available: http://www.cms.dmu.ac.uk/~it94jp/Project/Report/intro.htm
Ratner, J. A., Hyman, K., & Gondoly, K. (2000). Fun and fast testing methods for Web navigation. In *UPA 9th conference proceedings*. Asheville, NC: UPA Association.
Ratner, J. A., Hyman, K., & Klee, M. (1999, September). Blackout technique: A quick method for evaluating web page design. *Common Ground, 9* (3), 15–16.
Ratner, J. A., & Gondoly, K. (1998). Interactive design review process. In *UPA 8th conference proceedings*. Sedona, AZ: UPA Association.

III

Design and Evaluation Phases—Research in Ease-of-Use and Innovative Perspectives

8

Cognition and the Web: Moving From Theory to Web Design

Mary P. Czerwinski
Kevin Larson
Microsoft Research

THE WEAK CONNECTION BETWEEN COGNITIVE THEORY AND WEB DESIGN

Cognitive scientists who work in the Web design industry typically find themselves designing and evaluating complex Web-based systems to aid humans in a wide range of problem domains, like e-business, interpersonal communications, information access, remote meeting support, news reporting, or even gaming situations. In these domains, the technologies and the users' tasks are in a constant state of flux, evolution, and coevolution. Cognitive scientists working in human–computer interaction, and Web design may try to start from first principles developing these Web-based systems, but they often encounter novel usage scenarios for which no guidance is available (Dumais & Czerwinski, 2001).

For this reason, we believe that there is not as much application of theories, models, and specific findings from basic cognitive research to user interface design as one would hope. However, several analysis techniques and some guidelines generated from the literature are useful. In this chapter we outline some efforts in human–computer interaction (HCI) research from our own industrial research experience, demonstrating at what points in the Web design cycle HCI practitioners typically draw from the cognitive literature. We also point out areas where no ties exist. It is our goal to highlight opportunities for the two disciplines to work together to the mutual benefit of both.

In addition, our argument holds true for several other research areas, including both sociocultural and technological fields' theoretical and research contributions. We argue primarily from a cognitive viewpoint, as that is our background, and because the domain is resplendent with low-hanging fruit that could be applied to Website design.

Throughout our discussion, we provide examples from our own applied Web research, as well as from the work of others. Our goal will be to highlight where the discipline of cognitive science aided our design efforts, as well as where more effort needs to be applied before a thorough set of Web design practices and guidelines from any particular subfield of cognitive science can be instantiated.

PRACTICING WEB DESIGN WITH A STRONG COGNITIVE FOUNDATION

We cannot currently design complex Websites from first principles. Useful principles can be drawn from the subdomains of sensation, perception, attention, memory, and decision making to guide us on issues surrounding screen layout, information grouping, and menu length, depth, and breadth. Guidelines exist for how to use color, how to use animation and shading, or even what parameters influence immersion in Web-based virtual worlds.

Web designers also often borrow ideas from best practices, gleaned from very successful Websites that have been iteratively refined and accepted broadly in the electronic marketplace. However, there will always be technology-usage scenarios for which the basic research simply does not exist to guide us during design. It is well publicized that users scan Web pages, and most are not regular users of any one Website. These two facts alone make it fairly likely that undertaking good design for the Web will be daunting compared to that of more frequently used software applications.

In addition, Websites are used by a very wide range of users for a large variety of tasks—far more than the traditional office software application. Finally, the information that is on the Web is extremely complex material to work with, which makes the design task all the more challenging.

We feel that to move toward a stronger cognitive foundation in Web design, changes in the way basic research in cognitive science is performed would have to be put into place. Basic cognitive science research has not even begun to scratch the surface in terms of complexity in its paradigms, as is required when designing a Website today.

In addition, the breadth of users and the individual differences therefore encountered are equally not addressed in most psychological research in information processing. The stimuli and tasks typically used in cognitive research are too simple, the mental load is often not high enough, and the user is often not driven to perform based on emotional, real-world goals and constraints in the experimental scenarios used.

It is for this reason that many of the theories coming out of cognitive psychology (and many other subdisciplines in psychology) make predictions at far too low of a level in the knowledge worker's mental repertoire to be of much use to a Web designer. Although the cognitive information processing approach has valuable findings and methods to pull from, the approaches, theories, and findings will have to more closely approach realism to build a strong cognitive foundation in Web design practice. So, although there are a large number of findings within cognitive science from which Web designers can draw, knowledge falls far short of covering the entire practical needs of a Web designer. Next, we will review a few methods in the areas of cognitive science that have proven most useful to Web user interface design.

EXPERIMENTAL METHODS

Many of the methods that cognitive scientists use during their academic careers are very useful to Web practitioners and researchers. This is often the first way in which a cognitive scientist in industry can add value to a product team. The inventors have a particular design idea for a Website, but they typically have little experience in designing the right experiments and tasks to validate those design intuitions.

In our work, we have used visual search tasks, dual tasks, reaction time and accuracy studies, deadline procedures, memory methods like the cued recall task, and others to explore new Web technologies and interaction techniques. Problems with the proposed Website designs are almost always identified, as rarely is a new Web design flawless in its first stages. Also, iterative design, test, and redesign work to improve the existing technology implementation from that point forward.

The downside of using traditional experimental designs and tasks in Web work is that factorial designs with tight control and many subjects are simply not feasible given the time and resource constraints faced by design professionals. In addition, most real-world designs consist of many variables, and studying all possible combinations just isn't possible (or perhaps appropriate) given the short turnaround for which cycle Web design teams are notorious.

Finally, it is often the unanticipated uses of a Web site that are the most problematic, and these by their very nature are difficult to bring into the lab ahead of time. To understand some of the issues, think about how you would go about designing and evaluating a new voice-input Web portal over a 6-month period across multiple users, tasks, hardware systems, and languages.

One might conjecture that important independent variables could be studied in isolation, with the best of these being combined into a final design. In reality, this rarely works in Web design. We have witnessed teams that have iteratively and carefully tested individual features of a Web design until they were perfected only to see interactions and trade-offs appear when all of the features were united in the final stages of design or when the site was used in the real world.

Let's take a simple example of navigation bars. It may be that the labels and layout are intuitive and distinctive, as evidenced by user studies of early prototypes. However, when placed on the full Web page, complete with banner advertisements, notifications, and additional navigation content above, below, or to the side of the original navigation bar, users cannot find what they are looking for. Navigation content is easily swallowed when housed in the framework of the attention-grabbing content of most Web pages.

It takes careful iterative test and design of the full Web page, not just the single navigation bar, to get the visual layout of the bar to the level of priority necessary so that the user can complete their navigation tasks quickly and easily.

In addition to experimental methods, practitioners use a wide range of observational techniques and heuristics (e.g., field studies, contextual inquiry, heuristic evaluation, cognitive walk-through, rapid prototyping, questionnaires, focus groups, personas and scenarios, competitive benchmarks tests, usage log collection, etc.) to better understand Website usage and to inform design. These techniques are often borrowed from anthropology or sociology, and rarely would a Web design be complete without these research techniques and tools. Often a qualitative description of the jobs people are trying to do, or of how a Website is used in the field by real people doing their real work is much more valuable than a quantitative lab study of one small Website feature.

Observational techniques, task analysis, and related skills are much used in the practice of Web design but little represented in most cognitive science curricula. This is an area in which researchers in cognition find their background and training lacking when first introduced to applied work. In addition, the methods themselves are not necessarily as well honed or as useful as they could be in influencing Web user interface design.

Evaluating and iterating on existing Web user interfaces is only one aspect of the Web design task. Generating a good design in the first place or generating alternatives given initial user experiences is equally important. In order to do this, Web designers need to pull from basic principles in cognition, as well as basic principles of design.

Certainly, there are areas in which basic cognitive research has helped the domain of Web design by providing specific findings and guidelines for design. Some of the best-known and often cited examples of the applicability of results from basic psychology are the rather low-level, perceptual–motor findings that have been used quite effectively in Web design. For example, Fitt's law and Hick's law have been used to design and evaluate input device use during HCI for years.

The power law of practice and the known limits on auditory and visual perception have often been leveraged in the design of interactive Web systems. Many other results are seen in guidelines for good layout. For example, the use of color and highlighting, depth and breadth trade-offs in menu design and hyperlink grouping, and many general abstractions have made their way into "libraries" of parameters describing typical response times. Using findings and guidelines like these allow Web designers to start with a good initial design or to prevent silly mistakes, but it doesn't guarantee a useful, usable Website when all the variables are combined together into one design.

Even though specific findings and guidelines have proven useful in some cases, there also exist many problems in their use, which limit their effectiveness. Guidelines are often misused or misinterpreted, for example, using the rule of thumb of having only between seven plus or minus two items in a menu or on a Website, applying visual search guidelines but not taking into account the frequency with which items occur or the quality of category and label cohesiveness.

In addition, guidelines are often written at too abstract a level to help with specific designs, or, alternatively, they are too specific for a given usage context. A designer often finds it very difficult to combine all of the recommendations from specific findings or guidelines without knowledge of the costs, benefits, or trade-offs of doing so.

It is not our intent to state that there are no basic principles from which to design Websites available from theories in cognitive science; basic principles are available and do exist. We will highlight several of these principles below, noting when and where in the design cycle they were applied. Still, we argue that not enough of them exist and that more work with complex, user decision making, multitasking, and mental load needs to be carried out, in addition to working with complex information stimuli.

One specific area in which Web designers have been able to draw quite readily from cognitive theory is in the domain of visual attention. Especially with regard to the manner in which users will typically scan a Web page for content or banner ads, cognitive scientists have provided many great design principles.

For instance, Lim and Wogalter (2000) report two studies that looked at the placement of static banners on the Web. Recognition was shown to be reliably higher for banners in the top left or bottom right corners of the displays. They then extended this finding by placing banners in 16 regions across the display—outer, intermediate, and central regions. Each region included four corners of banner placement.

Results showed that recognition performance was reliably higher for banners centrally located over those in the outer regions of the display. The intermediate regions were not significantly different from the outer or inner regions. A separate analysis was carried out on the four corners for each region, and again the top left and bottom right corners were significantly better recognized than the other two corners. The results suggest that advertisers would do well to place their banner ads spatially on a continuum from top left, middle, or bottom right of the display rather than in other areas of the layout. In addition, the authors argue that "visual pagers," or notifications, could be made more salient by using this spatial location positioning. However, they warn that these notifications could also be very distracting in these locations, and so they caution this placement advice for tasks that rely on the user giving their full attention to the primary task (e.g., monitoring tasks like air traffic control).

Visual attention theorists have also provided us with good examples for presenting Web content in a way that will grab users' attention without much effort—called preattentive processing by Treisman (1985). Typically, tasks that can be performed on large multielement displays in less than 200 to 250 msec are considered preattentive. Eye movements take at least 200 msec to initiate, and random locations of the elements in the display ensure that attention cannot be prefocused on any particular location, yet subjects report that these tasks can be completed with very little effort. This suggests that certain information in the display is processed in parallel by the low-level visual system. One very interesting result has been the discovery of a limited set of visual properties that are processed preattentively without the need for focused attention. Visual properties that are processed preattentively can be used to highlight important Web display characteristics.

The kinds of features that can be processed in this manner include line orientation, length, size, number, terminators, intersections, closure, color flicker, and direction of motion, as well as a few newly discovered 3-D cues. Experiments in cognitive science have utilized preattentive features to show that users can rapidly and accurately detect the presence or absence of a "target" with a unique visual feature within a field of distracters, can rapidly and accurately detect a texture boundary between two groups of items where all of the elements in each group have a common visual property, and can count or estimate the number of elements in a display with a unique visual feature. Obviously, you can't leverage all of these preattentive cues and features at once, and research is starting to show that some of these preattentive features take precedence over others.

For instance, Callaghan (1990) has reported studies that suggest that brightness overrides hue information and that hue overrides shape during boundary detection tasks. However, knowing what features can work toward popping out various sections of your Web display can give the designer an added advantage if warranted for a particularly important user task. To exemplify this, let's consider Fig. 8.1, a snapshot of http://www.corbis.com. On the screen, color is used to pop out the categories of interest, and very little information is competing with the viewer's attention to this area of the display. In addition, the image selected for the home page has a background that blends in nicely with the chosen links for navigation, adding a sense of symmetry and grouping to the pop-out effect.

Likewise, there is much we know from the early work of the Gestalt psychologists about good Web design. The Gestalt theorists gave us the principles of proximity, similarity, common fate, good continuation, and closure. Their "Laws of Organization" primarily focused on recognizing form from background. For instance, the principle of area states that the smaller a closed area, the more it is seen as the figure. The principle of proximity states that objects close to each other tend to be grouped together. The principle of closedness states that areas with

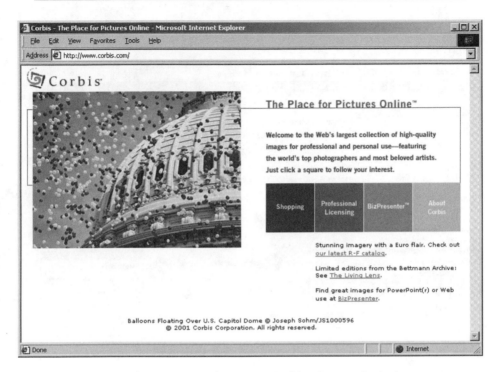

FIG. 8.1. http://www.corbis.com, Corbis Corporation's home page with color boxes used for categories of interest, as well as minimal words on the page.

closed contours tend to be seen as figure more than do those with open contours. The law of symmetry holds that the more symmetrical a closed region, the more it tends to be seen as figure. The law of good continuation states that the particular arrangement of figure and ground that is seen is the one that requires the fewest changes in straight or smoothly curving lines or contours.

All of these principles can easily be put into place during good Web design. By way of example, let's examine Fig. 8.2, a screen shot of Fidelity's Website. This Website was iteratively redesigned after studies revealed that users were ignoring areas of the site thought to be of core importance (Tullis, 1998). It was thought that layouts that "guided the eyes" to more important parts of the site would generate more customers to key areas. Note the great use of grouping and symmetry in the Web page's layout, resulting in six distinct groups (including the navigation bar at the top) balanced from left to right across the page. Note, too, how the new user is roped into getting started by a prominent "New to Fidelity?" link squarely centered on the page, foveally, in an area most likely to get scanned. The Website is leveraging visual perception and attention principles in a very strong way.

Coming out of the visual design field, Bertin (1983) proposed the "semiology of graphics" and systematically classified the use of visual elements to display data

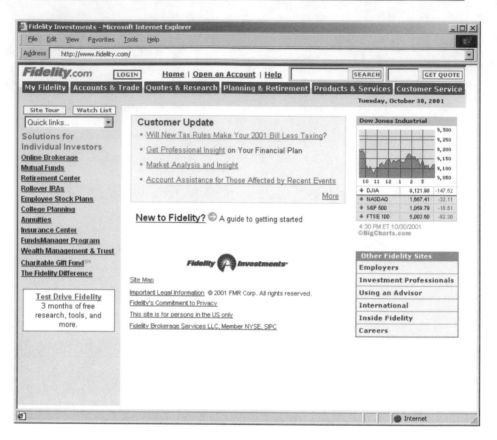

FIG. 8.2.   The Fidelity.com Website (www.fidelity.com).

and relationships. His system consisted of seven visual variables: position, form, orientation, color, texture, value, and size. He joined these visual variables with visual semantics for linking data attributes to visual elements. Note the similarity between Bertin's seven visual variables and the preattentive features proposed by cognitive researchers. Clearly this is one place were Web designers can benefit from principles converging from two completely separate disciplines.

Spatial layout can be successfully applied to Web pages in order to effectively leverage human spatial memory, as has been demonstrated quite effectively over the last few years (Chen & Czerwinski, 1997; Czerwinski et al., 1999; Robertson et al., 1998; Tavanti & Lind, 2001). Using spatial layout of your Website for this purpose will ensure that users will develop a better conceptual model of your site's content over fewer return visits. Because a user's willingness to return to a site hinges on its ease of use and navigability (Lohse & Spiller, 1998), spatial memory for the site structure may be key. Research by architects has shown that distinctive paths, nodes, landmarks, boundaries and districts are very important to leverage this remarkable human capability (Lynch, 1960; Passini, 1984). Vinson (1999), in particular, has laid out very clear guidelines for user interface design with good landmarks for navigation. He stresses the importance of using landmarks that are distinctive, fall on clear gridlines with proper axis alignment, and scale appropriately for the world in which they are placed.

In addition to the use of landmarks, we have found in our research that allowing the user to place content where they want it further enhances spatial memory, likely because the further processing that is required (both motorically through manual placement on the page and cognitively in terms of deciding where the content should go). Of course there is a trade-off here in terms of whether or not users will tolerate personalizing the Web page, because this requires additional effort. Still, if the user is bound to return multiple times, the effort may outweigh the cost of trying to navigate a site that is suboptimally designed for a particular user's task. We think the Austin MONKEYmedia Website could demonstrate an effective use of landmarks to enhance navigability of their site, except for the simple problem that the landmarks shift spatially as the user moves from subsection to subsection of the site. A snapshot of their home page is shown in Fig. 8.3.

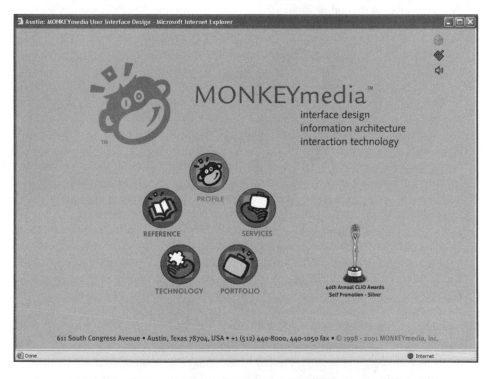

FIG. 8.3. The MONKEYmedia Website (the ultrasensory Website available at http://www.monkey.com/FR4/main.htm).

The Website uses a set of five clearly distinctive landmarks to help the user navigate through these cornerstones of the site from any location. Interesting animation and audio effects underline the distinctive landmarks, although it is unclear if these additional cues are useful.

Well-known psychological principles do not always apply easily. Larson and Czerwinski (1998) provided an example analyzing use of the rule of thumb recommending seven plus or minus two links on a Website. This guideline germinates from a famous paper by Miller (1956), where the "magic number" describes the number of "chunks" of information an average person can hold temporarily in short-term memory. In several Websites that we worked on, we observed that if the number of links at the top level of the site exceeded a small number, novice Web users made more errors when trying to find particular content. We speculated that short-term memory was a factor in how many links can be presented on a page for optimal access to a document and that the number of links should not exceed seven plus or minus two.

To test this hypothesis, we created three Websites with 512 documents that were similar in content but with varying structure in the number of links between which users would have to choose. One Website presented three levels of links with 8 links per level, a second presented 16 at the top level and 32 at the second level, and a third site had 32 at the top level and 16 at the second level. It turned out that users performed reliably slowest and were most lost when the number of links per level was held to 8. Users were fastest and least lost when the Website contained 16 top-level items, but this site was not reliably faster than the Website with 32 top-level items.

In Miller's (1956) famous article about the limits of short-term memory, he reported a study in which subjects were asked to try to match a new tone to a set of existing tones. If the set of existing tones is small, subjects are very proficient at the task but perform much worse as the set grows. Superficially picking between tones and picking between a list of links seems very similar. Our study suggested that these are very different tasks. When picking between Web links, subject's short-term memory does not appear to be an important factor. Instead, it is far more important to ensure that the labeling is good enough to tell people what information will be available if they follow the link.

Indeed, recent work by Dumais and others (Chen & Dumais, 2000; Dumais, Cutrell, & Chen, 2001) reports that the good visual categorization (spatial grouping and labeling) of Web content is key to optimizing the presentation of navigable search results. An example applying this idea (one of the designs that tested successfully in the series of studies carried out by Dumais et al. (2000; 2001)) is presented in Fig. 8.4.

One new phenomenon recently identified by psychologists that directly applies to Web design is what is referred to as "change blindness" (Benway, 1998). Change blindness is a perceptual and cognitive phenomenon that occurs when movement that typically accompanies change is masked. The masking could be

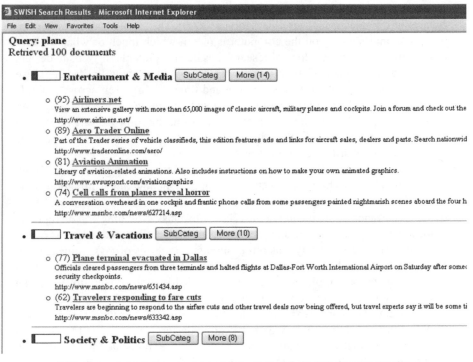

FIG. 8.4.   A partial snapshot of a successful design for presenting
Web search results (adapted from Dumais, Cutrell, & Chen, 2001).

brought about by blinking or being momentarily distracted by something elsewhere in the visual field. Change blindness occurs when the user does not notice a
difference to the visual information from before to after the attentional distraction.
Psychologists have verified that this change blindness occurs both in the real
world and in graphic displays, and that a visual disruption as brief at 80 to
240 msec can result in a user not noticing a change in the display (see the Website
at http://www.acm.org/sigchi/bulletin/2001.6 for an interesting example of this
phenomenon).

   What this implies for Web designers is that if your users are traversing pages that
are related to each other and that only subtle differences need to be identified by the
user, the download time between navigating from one page to another may result
in the user not noticing that they've arrived at the new page. In addition, as pointed
out by Benway (1998), directing the user to problems during form validation is a
ripe area for change blindness to be exhibited. The designer will need to come up
with attention attracting visual (and maybe audio) cues to direct the user to those
fields in the form that require the user's attention or else change blindness may
result.

Another example from our work in managing interruptions illustrates some of these issues (Czerwinski, Cutrell, & Horvitz, 2000). The problem area of interruption while multitasking on the computer is one in which much of what we have learned from basic psychological research has been applicable in Website design. For example, 100 years of attention research has taught us about limited cognitive resources and the costs of task switching and time sharing even across multiple perceptual channels. But how does one go about designing an effective Web user interface for alerts when the user may benefit from the information? Also, how does one design an interface that allows the user to easily get back to the primary Web task after a disruptive notification?

In practice, specific findings guide our intuitions that flashing, moving, abrupt onset, and loud audio heralds will attract attention. But how much attraction is too much? How does this change over time? Do users habituate to the alerts? How does the relevance of the incoming messages affect task switching and disruption? Some studies in the literature suggested relevance was influential, whereas others found that surface similarity was more important in terms of task influence. All of these studies used tasks that were not representative of typical computer tasks (i.e., they were too simple or demanded equal distribution of attention). We ran our own studies and showed the benefits of only displaying relevant notifications to the current task in order to mitigate deleterious effects (Czerwinski et al., 2000).

When, as part of the same Web notification design, we attempted to design a "reminder" cue that was visual and could help the user reinstate the original task context after an incoming notification, specific findings again were not useful. We found through our own lab studies that using a "visual marker" as a spatial place-holder to get users back to a point in a primary Web task was not enough. Instead, users needed a "cognitive marker" that provided more of the contextual, semantic cues related to a primary task (Cutrell et al., 2001). No specific findings existed in the literature for exactly what visual anchors could be used in a Web display to help users tap into their mental representations of a task or how many visual retrieval cues would be needed to reinstate a Web task context most efficiently. Again, prototypes needed to be designed and tested.

Our studies over 2 years taught us many important principles in the design of notifications to computing systems. Most importantly, the system must learn some information about the user's current context (e.g., how busy is the user? What applications are being used? What is the topic of focus? How important is the incoming information? How relevant is it to the current topic of focus or one that is coming up shortly? What is the best way to present the information to the user given his or her context? On what device?). Though there are obvious privacy concerns here, we have found that users are mostly enthusiastic about a system that learns their habits over time and respects their current level of "interruptability."

We found in several studies that a good notification system should wait for a motoric or cognitive "break" before presenting information to the user (Czerwinski et al., 2000) and that information that is relevant to the user's current task will be easier to attend to and then dismiss in order to return to the primary task (Czerwinski et al., 2000). Studies also showed that users were more likely to forget their primary task content if interruptions or notifications came earlier into the task as opposed to later (Cutrell et al., 2001). We found that users were not be likely to trust a faulty intelligent system, so it is very important that an intelligent notification system has good default behavior from the onset (Tiernan, Czerwinski, & Cutrell, 2001). Finally, our research showed that repeating information via notification that a user has either already seen or that is now irrelevant to ongoing tasks is significantly more frustrating than a first or timely delivery of the same information.

In both the design of Web notifications and the corresponding Web reminders, paradigms from very basic cognitive research had to be utilized in new lab studies to examine the broader usage context and the specific user interface design chosen for the target task domains.

NEW WEB METRICS: SUBJECTIVE DURATION ASSESSMENT

Because of the lack of fit between Web design practice and findings in cognitive science, our basic research also attempts to develop new metrics for Web interaction assessment. Our lab has explored a new approach we refer to as *subjective duration assessment* for gauging users' difficulties with Web tasks, interfaces, and situations (Czerwinski et al., 2001). The approach, adapted from a finding described in the interruption literature in psychology, centers on the use of time estimation to characterize performance. We introduce a metric, named *relative subjective duration* (RSD), which provides a means for probing the difficulty that users have with performing Web tasks—without directly asking users about the difficulty and invoking any demand characteristics that explicit probing might engender. Although there are several applications of RSD, we focused on the use of this measure to probe users' experiences with new Website designs without directly asking them for feedback. Direct assessment of satisfaction has been found to be frequently confounded by an inherent bias toward the positive end of the scale.

RSD is based on a discovery nearly 75 years ago showing that when engaging tasks are interrupted, participants tend to overestimate how long those tasks take when compared to actual task times. Conversely, tasks that are completed tend to be underestimated in terms of the overall task times. We explored the value of time estimation as a metric for evaluating task performance in Web design. Our hypothesis was that participants would overestimate the duration of activity on tasks that are halted before completion because users were not able to complete them on their own. In contrast, we believed users would underestimate the duration of activity on tasks completed successfully.

A user study of a well-known Internet browser explored the efficacy of the time estimation metric against both satisfaction and actual performance metrics during common Web tasks. Our results showed that using subjective duration assessment can be a valuable tool for Web user interface research. It circumvents the need to explicitly ask users for satisfaction ratings and therefore is a better, implicit measure of Web satisfaction that is more positively correlated with actual performance metrics. We hope to see more novel Web interaction metrics emerge from laboratories doing advanced technology development over the years to come.

BEYOND THE DESKTOP—FUTURE WEB INTERFACES

We conclude this chapter with a selective glimpse at advanced Web technology from our labs. The projects described below are currently being developed for deployment on the Web over the next 2 to 5 years and are targeted toward making the Web more useful for the mobile Web user. Note that this technology is still so new and evolving so rapidly that it is yet unclear whether or how basic cognitive principles should or could be applied. Our approach has been to use a combination of iterative test and design in parallel with basic research on human interaction in order to shed more light on these matters. First we will discuss research on an audio feedback mechanism that will be useful for today's Websites but will become more important for supporting mobile users. Then we will discuss current projects for supporting access to the Web over telephony and multimodal devices.

Nonspeech Audio

Ease of navigation is one of the most important elements of a good Website, and the only tool we currently use to support navigation is visual feedback. We attempted to augment visual feedback with nonspeech audio feedback on one of our Websites. In previous work with this site we learned that people tended to get stuck between the second and third levels of this site when looking for information and would forget to research the home page.

We wanted to use audio in an unobtrusive way to let people know how deep they were in the Website. Users didn't need to do anything extra to hear the signal; it would play whenever a link was selected. There were two dimensions coded in the signal. The first dimension of the signal provided depth information by pulsing the number of levels of depth of the selected item; the top-level received one pulse, the second-level item received two pulses, and the third-level item received three pulses. The second dimension coded into the audio was direction of movement; selecting an item on a lower level elicited a decreasing pitch, whereas selecting an item on a higher level elicited an increasing pitch. It was expected that users would use this information to help navigate around the hierarchy and to not get trapped in an incorrect subsection of the site.

Unfortunately, there was no indication that the auditory feedback improved performance in the search task; the reaction times for finding particular items in the Website were not statistically different between the nonaugmented site and when the audio feedback was added to the site. Similarly, the users did not reliably rate the Website more appealing when the audio feedback was present.

In future studies, we plan to examine different kinds of nonspeech audio to try to make the auditory feedback more understandable and to also examine increased amounts of experience with the audio feedback. Although nonspeech audio on desktop devices will be redundant with visual feedback, it will become a crucial form of feedback when we try to make the Web accessible over a telephone.

Telephony

WIMP (windows, icons, menus, and pointers) interfaces on desktop computers have been very successful and are the most frequent way people currently access the Web. In the future, there will be a greater variety of devices used to connect to the Web, and the desktop input techniques will not scale well. One such device is the common phone.

The phone is by far the world's most ubiquitous communication device, and opening up the Web to anyone with a phone greatly increases the number of people who will have access to the Web. Most phones do not support WIMP interfaces. Content must be accessed with either key presses or speech recognition, and content must be displayed with either recorded speech for static content or text-to-speech (TTS). There are already a couple of commercial services for making Web content available over the phone and others that are currently under development.

Recently, we compared three telephony interfaces that allow users to access stock reports, weather reports, and personal information content like e-mail. These interfaces worked by presenting the user with a list of the services that the system provided, and the user would select the service by saying its name. Once in the system, the appropriate features would be described, and the user would again chose between options until the user drilled down far enough to reach the target information. Users were very successful with all three interfaces at accessing information like the local weather or the stock price for a particular company, but e-mail systems were not as successful. Users had no problem navigating through their e-mail box with commands like next and read, but these commands were not sufficient for navigating though more than a small handful of messages.

There are several challenges for accessing the Web over the phone. The current state of TTS technology is that it is good enough for people to clearly recognize all the words said but that it does not sound natural enough for people to listen to for long periods of time. Second, telephony interfaces need to allow users the flexibility to change their mind at any time. Users expect to be able to stop the system at any time to do something different, and the systems currently are not always good at allowing barge-in. Barge-in will have to be universally available, and feedback on the new action must be quick. Third, telephony systems need to figure out how to present data that are more complex. This is one area in which nonspeech audio can help, but the content itself will have to support the user's tasks. Reading through 5 e-mails one at a time is possible, but users do not want to read 50 e-mails like that. Users will not read e-mail over the phone in the same way that they do on WIMP interfaces, and we need to consider how their usage will differ. Task analysis will take on even greater importance for these devices.

Next, we turn to an exciting new kind of interface that is coming a little after telephone portals. These interfaces will present information on a display and allow the user to use speech for navigation and text entry.

Speech and Mobile

For the Web to become truly ubiquitous, it must be accessible at anytime from anyplace. The current generation of PDAs (Personal Digital Assistants: Palm and PocketPC devices) are the first computing devices that are truly portable enough to bring anywhere. Unfortunately, these devices provide an unsatisfactory Web experience because navigation and text entry are much too difficult to access on a small screen and without a keyboard. MiPad, a prototype used in our labs, uses a combination of speech recognition and natural language capabilities to allow users to be more productive with PDAs (Hwang et al., 2000). MiPad lets users issue commands like "Schedule a 30-minute meeting with Bill Gates for Friday at 2 in the afternoon," which will cause a new appointment to be opened and all the appropriate fields filled in. Users can also write new e-mail messages with MiPad by dictating the text they want sent. User studies show that users can schedule appointments and transcribe e-mails with the MiPad speech interface reliably faster than with a soft keyboard on the same PDA.

A key challenge here is correcting speech recognition errors. Speech recognition accuracy is in the neighborhood of 95%, which means that 1 word out of every 20 needs to be corrected. It is this correction time that keeps speech recognition from being more efficient than typing on a desktop machine. We need to develop strategies for reducing the time for correction. In a recent study in our labs we found that using speech-only to correct speech errors is terribly inefficient because it leads to many additional errors. Using either an alternates list to correct errors or a multimodal correction technique with a pen to select and speech to redictate is better than speech only.

Recently, we have turned our attention from developing applications that use speech on a PDA to developing the framework that will enable anyone to turn their regular Websites into speech-enabled Websites. Developers will be able to write Speech Application Language Tags (SALT; www.saltforum.org) to let people interact with Websites using speech. SALT can work on any device from a simple phone to a desktop, but certainly, its sweet spot is enabling mobile interaction. Being able to ask a Website for flight times from Seattle to New York and seeing a list of flights is better than either using a phone and hearing a list of flights or having to use inefficient input methods on the PDA.

In addition to using SALT for form filling on the Web, if combined with natural language technologies, users will be able to simply ask for information and have it automatically delivered to them rather than being required to perform traditional navigation on a small screen. As we move forward with this we will face the challenges of efficiently correcting the speech recognition errors and developing the natural language interface that will be used to navigate both between and within Websites.

CONCLUSION

We opened this chapter with the premise that the design and evaluation of Web sites is challenging. We further argue that basic results and theory can provide reasonable starting places for design, although experience in an application domain is perhaps as good. Theories and findings from psychology are not, however, as useful as one might hope. Getting the design right from the beginning is very hard even with these starting points, so any real-world design will involve an ongoing cycle of design, evaluation, failure analysis, and redesign.

How do we move from the current state of affairs to a better symbiosis between basic research in cognitive science and Web design to eradicate overreliance on Web designs that are "point" designs or the examination of Web features in complete isolation for simplistic tasks? We would like to propose moving basic research in cognitive science closer to the real needs of Web designers and human–computer practitioners by studying more complex, real-world problem domains. Examples ripe for investigation include such popular Web domains as purchasing items on the Web and searching for information. We do, however, feel as though progress is being made. For instance, starting with the variables discussed in this chapter (short-term memory, visual attention dimensions, notifications and task switching, information retrieval and speech), the Web practitioner can begin to see basic guidelines for design emerging from carefully controlled studies of the intended Web behavior. It is our vision that increasingly more academic and industrial research partners will work toward common goals in building a better foundation for Web design.

REFERENCES

Benway, J. P. (1998). Banner blindness: the irony of attention grabbing on the World Wide Web. In Proceedings of the Human Factors and Ergonomics Society 42nd Annual Meeting, Chicago, Vol. 1, pp. 463–467.

Bertin, J. (1981). *Graphics and graphic information processing*. Berlin: Walter de Gruyter & Company.

Callaghan, T. C. (1990). Interference and Dominance in Texture Segregation. In D. Brogan (Ed.), *Visual search* (pp. 81–87). New York: Taylor & Francis.

Chen, C., & Czerwinski, M. (1997). Spatial ability and visual navigation: an empirical study. In The New Review of Hypermedia and Multimedia (Vol. 3, pp. 67–91). London: Taylor Graham.

Chen, H., & Dumais, S. T. (2000). Bringing order to the web: Automatically categorizing search results. In *Proceedings of Association for Computing Machinery's CHI 2000: Human factors in computing systems* (pp. 145–152). New York: ACM Press.

Cutrell, E., Czerwinski, M., & Horvitz, E. (2001). Notification, disruption and memory: Effects of messaging interruptions on memory and performance. In M. Hirose (Ed.), *Proceedings of Human–Computer Interaction—Interact 2001* (pp. 263–269). Tokyo: IOS Press.

Czerwinski, M., Cutrell, E., & Horvitz, E. (2000). Instant messaging: effects of relevance and time. In S. Turner & P. Turner (Eds.), *People and computers: Vol. 14. Proceedings of HCI '00* (Vol. 2, pp. 71–76). Edinburgh British Computer Society.

Czerwinski, M., Horvitz, E., & Cutrell, E. (2001). Subjective duration assessment: An implicit probe for software usability. In D. Benyon & P. Palanque (Eds.), *People and Computers: Vol. 15. Proceedings of HCI '01* (Vol. 2, pp. 167). Lille: British Computer Society.

Czerwinski, M., van Dantzich, M., Robertson, G. G., & Hoffman, H. (1999). The contribution of thumbnail image, mouse-over text and spatial location memory to Web page retrieval in 3D. In A. Sasse & C. Johnson (Eds.), *Human–Computer Interaction—Proceedings of Interact '99* (pp. 163–170). Edinburgh, Scotland: IOS Press.

Dumais, S. T., Cutrell, E., & Chen, H. (2001). Optimizing search by showing results in context. In *Proceedings of Association for Computing Machinery's CHI 2001: Human factors in computing systems* (pp. 277–283). New York: ACM Press.

Dumais, S. T., & Czerwinski, M. (2001, August). Building bridges from theory to practice. In Usability evaluation and Interface design: Proceedings of *HCI International 2001* (Vol. 1, pp. 1358–1362). Mahwah, NJ: LEA.

Fitts, P. M. (1954). The information capacity of the human motor system in controlling the amplitude of movement. *Journal of Experimental Psychology, 47,* 381–391.

Hick, W. E. (1952). On the rate of gain of information. *Quarterly Journal of Experimental Psychology, 4,* 11–26.

Hwang, X., Acero, A., Chelba, C., Deng, L., Duchene, D., Goodman, J., Hon, H., Jacoby, D., Jiang, L., Loynd, R., Mahajan, M., Mau, P., Meredith, S., Mughal, S., Neto, S., Plumpe, M., Steury, K., Venolia, G., Wang, K., Wang, Y. (2001). Mipad: A multimodal interactive prototype. In *Proceedings of the IEEE International Conference on Acoustics, Speech, and Signal Processing* (Vol. 1, pp. 9–12). Piscataway: IEEE Computer Society.

Larson, K., & Czerwinski, M. (1998). Web page design: Implications of memory, structure and scent for information retrieval. In *Proceedings of Association for Computing Machinery's CHI '98, Human Factors in Computing Systems* (pp. 25–32). New York: ACM Press.

Lim, R. W., & Wogalter, M. S. (2000). The position of static and on–off banners in WWW displays on subsequent recognition. In *Proceedings of the IEA 2000/HFES 2000 Congress* (pp. 420–423). Santa Monica, CA: Human Factors and Ergonomics Society.

Lohse, G. L., & Spiller, P. (1998). Quantifying the effect of user interface design features on cyberstore traffic and sales. In *Proceedings of Association for Computing Machinery's CHI '98: Human Factors in Computing Systems* (pp. 211–218). New York: ACM Press.

Lynch, K. (1960). *The image of the city.* Cambridge, MA: MIT Press.

Miller, G. (1956). The magical number seven, plus or minus two: Some limits on our capacity for processing information. *The Psychological Review* (Vol. 63, pp. 81–97). Washington, DC: APA Journals.

Passini, R. (1984). *Wayfinding in architecture.* New York: Van Nostrand Reinhold.

Robertson, G., Czerwinski, M., Larson, K., Robbins, D., Thiel, D., & van Dantzich, M. (1998). Data mountain: Using spatial memory for document management. In *Proceedings of UIST '98, 11th*

Annual Symposium on User Interface Software and Technology (pp. 153–162). New York: ACM Press.

Tavanti, M., & Lind, M. (2001). 2D vs. 3D, Implications on spatial memory. In *Proceedings of IEEE's Information Visualization* (pp. 139–145). Los Alamitos, CA: IEEE Computer Society.

Tiernan, S., Czerwinski, M., & Cutrell, E. (2001). Effective notification systems depend on user trust. In M. Hirose (Ed.), *Proceedings of Human–Computer Interaction—Interact 2001* (pp. 684–685). Tokyo: IOS Press.

Triesman, A. (1985). Preattentive processing in vision. In *Computer Vision, Graphics, and Image Processing* (Vol. 31, pp. 156–177).

Tullis, T. S. (1998). A method for evaluating web page design concepts. In *Proceedings of Association for Computing Machinery's CHI '98: Human factors in computing systems* (pp. 323–324). New York: ACM Press.

Vinson, N. G. (1999). Design guidelines for landmarks to support navigation in virtual environments. In *Proceedings of Association for Computing Machinery's CHI '99: Human factors in computing systems* (pp. 278–285). New York: ACM Press.

9

Designing Improved Error Messages for Web Browsers

Jonathan Lazar*
Yan Huang†
Towson University

INTRODUCTION

Errors frequently occur when users interact with computer systems, especially networked systems such as the Internet. When errors occur, users are typically presented with an error message. Unfortunately, many error messages are not helpful to the user and instead only confuse the user. In an ideal world, a user could interact with a computer system without any errors occurring. Realistically, errors occur, and the interface must provide a method for dealing with the errors. When an error does occur, the user is frequently presented with an error message. The error message informs the user that an error occurred. The error message should hopefully help the user deal with the situation and recover from the error sequence. Unfortunately, many error messages are confusing and hard for users to understand. Although usability has improved for many computer interfaces, error messages

* Jonathan Lazar is a faculty member in the Computer & Information Sciences Department at Towson University. He is also an affiliate Professor of Towson University's Center for Applied Information Technology.

† Yan Huang completed this research project when she was affiliated with Towson University. She is currently a senior user systems specialist at Johns Hopkins University, Baltimore, Maryland.

167

have not gained much attention in the past. However, this situation is changing. At the Association for Computing Machinery's annual CHI (Computer–Human Interaction) conference in 2001, the keynote speaker, Bill Gates, joked around that he was unable to understand one of Microsoft's error messages and asked the assembled crowd for suggestions. The well-known expert on Web usability, Jakob Nielsen, recently dedicated one of his "Alertbox" columns to the issue of customized error messages from Websites (Nielsen, 2001). Error messages were in the news recently because of incidents involving the GMAT test, the exam that is required for those who want to enter graduate business programs. For thousands of those taking the GMAT test on computer, a cryptic error message appeared at the end of the test, causing confusion and concern (Cable News Network, 2001).

Web browsers are one of the most popular software applications. Many new users have purchased a computer specifically to be able to access Websites. But Web browsers are increasingly the application of choice for many tasks, not only the browsing of Websites. Recent operating systems use the Web browser as a tool to explore the files on the user's computer. In addition, many corporate and organizational applications are being deployed through Web browsers. Corporate intranets are allowing users to access many different corporate information systems by using the Web browser as a tool that crosses platforms (Jacko et al., 2001). Application service providers are also selling access to software applications that are deployed through Web browsers (Hoffer, George, & Valacich, 2001). Because the Web browser is being used for an increasing number of tasks, the usability of the Web browser becomes increasingly important. Much as Websites should be developed based around the needs of the users, Web browsers should also be built to meet the needs of the users (Lazar, 2001).

RESEARCH ON ERROR MESSAGES

Interestingly enough, much of the research on error messages comes from the historic beginnings of the field of human–computer interaction. This research forms the foundations of what we know about error messages today. Norman suggests that there are three goals for system design related to error: (1) minimize the root causes of error, (2) make it easy for users to reverse their actions, and (3) make it easy to discover errors and then correct the problem (Norman, 1988). Error messages address the last goal, making it easy to discover errors and correct the problem.

Shneiderman (1998) suggests that error messages should state specifically what occurred (in language that the user can understand), be positive, and offer suggestions on how to respond. However, most software applications and operating systems unfortunately do not follow even the most basic guidelines that Shneiderman developed for error messages. Better error messages were discovered to improve user performance in a number of tasks (Shneiderman, 1982). Unfortunately, many current software applications still use terms such as *fatal* and *illegal* in their error messages. An error message saying something along the lines of "A fatal exception OE has occurred at 017F: BFFAADOB. The current application will be terminated" will in no way assist the user in responding to the error situation. In fact, this type of error message will likely confuse and frustrate the user. The user will not understand what has occurred and will not be able to respond in an appropriate manner. It is possible that the user will perform an inappropriate action, and instead of successfully exiting from the error sequence, a more serious error may occur, which is harder to recover from (Carroll & Carrithers, 1984).

A new trend in the human–computer interaction community is the topic of universal usability (Shneiderman, 2000). Universal usability refers to a push for information systems that are easy to use for all user populations. This includes novice users, expert users, users with different educational and economic levels, users with disabilities, and users with different browser brands and versions. Error messages that are user friendly can make a large leap forward toward the goal of universal usability. Although expert users may be able to overcome such an error situation, error messages that confuse and frustrate the user do not allow for an enjoyable user experience for novice and inexperienced users. Better error messages could assist a wide range of user populations with their interaction experience.

The Possibility of Errors on the Web

Errors are always a possibility when users are interacting with computer technology. There are many possible definitions of an error. For instance, from a user's point of view, an error could be considered to be a situation when users have the functionality available in a software application but are unable to reach their task goal (Lazar & Norcio, 2000). Another possibility is a more technically oriented view of error, which is that a user's command does not follow correct guidelines (Lazar & Norcio, 2000). Of course, blame shouldn't be an issue in an error; the goal should be to assist users in existing from an error sequence. For the purposes of this experiment, an error is defined as a situation in which an error message is displayed.

In the network environment, the chance for error increases. With all of the networks, routers, and servers involved in a simple Web page request, there is a likelihood of one of these components not working properly or failing (Lazar & Norcio, 2000). Although the user may have performed all of their tasks in a correct manner, he or she may still not be able to reach the task goal through no fault of their own (Lazar & Norcio, 2000). Norman separates errors into two categories: a *slip* and a *mistake* (Norman, 1983). A mistake is when a user chooses the wrong commands to reach a task goal (Norman, 1983). A slip is when a user chooses the correct commands to reach a task goal but implements the commands incorrectly (such as a spelling error, Norman, 1983). When a user has chosen the correct commands to reach the task goal, has entered those commands correctly, cannot reach the goal yet has no control over the situation, this is called a situational error (Lazar & Norcio, 2000). In most cases, even with a situational error, an error message will appear. However, it may not be clear in the error message whether the user can make any changes that will allow them to reach the task goal or whether the error is a situational error, in which the user may not be able to reach the task goal.

ERROR MESSAGES IN WEB BROWSERS

There are certainly many possible causes of error. In addition, there is the possibility that users may perceive that an error has occurred, although no incorrect actions have taken place. To assist in developing our research methodology, we examined some of the most commonly occurring error messages provided by the Web browser to determine the cause of the error message displayed. Individual web sites may also provide customized error messages. However, these messages are only displayed in limited circumstances and these messages are inconsistent from web site to web site. And while Internet Explorer's error messages are constantly changing (in fact, Internet Explorer's error message recently changed to provide a search engine, Olsen, 2001), Netscape Navigator's error messages have stayed constant from previous versions, through the new version, 6.0. For our study, we therefore identified seven specific error messages provided by Netscape Navigator. They are not listed in any specific order.

Error Message 1:
"This program has performed an illegal operation and will be shut down. If the problem persists, contact the program vendor."

The cause of the error:
The browser application has crashed. This could be due to an inappropriate request. A Java applet might also have crashed the browser application.

Error Message 2:
"Netscape is unable to locate the server: web.towson.edu. Please check the server name and try again" (see Fig. 9.1).

Note. The URL provided is a sample URL. The error message typically specifies whichever URL was requested.

The cause of the error:
The local domain name server was unable to find any Web servers matching the user's request. Most likely, the domain name (in this case, towson.edu) is correct, but the server name (web.towson.edu) is incorrect. It could also be possible that the Web server is having problems and is currently unavailable.

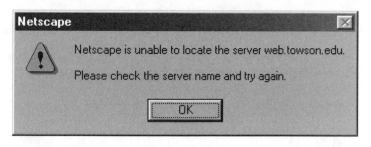

FIG. 9.1. An example of a common error message in Netscape
Navigator.

Error Message 3:
"Forbidden—You don't have permission to access www.towson.edu/~jlazar/
forbidden.html on this server."

Note. The URL provided is a sample URL. The error message typically specifies
whichever URL was requested.

The cause of the error:
The server understood the request but is refusing to fulfill it. The file protections
have not been set on the Web server to allow for full public access for the requested
files.

Error Message 4:
"Authorized failed. Retry?"

The cause of the error:
The URL request requires user authentication. The user typically must provide
a valid user name and password. Therefore, the user should attempt the request
again and provide an accurate user name and password.

Error Message 5:
"The web server can not find the file or script you asked for. Please check the URL
to ensure that the path is correct. Please contact the server's administrator if this
problem persists."

The cause of the error:
Most likely, the user provided the correct server name but an incorrect file name.

Error Message 6:
"There was no response. The server could be down or is not responding. If you
unable to connect again later, contact the server's administrator."

The cause of the error:
The Web server did not provide a timely response to the user's request. This could
be due to the Web server receiving a large number of transactions, or the Web
server itself might be down.

Error Message 7:
"Netscape is unable to locate the server. The server does not have a DNS entry. Check the server name in the location (URL) and try again."

The cause of the error:
This error message would most likely appear if there is a local domain name server that is not functioning properly. This is out of the user's control. This could also be due to the network connection being severed, which again is out of the user's control. It is also possible that this error is due to a user entering a totally incorrect domain name.

RESEARCH METHODOLOGY

This research was exploratory, with two major goals. One goal was to explore how to improve the working of current error messages. A second goal was to determine if users would respond differently to the new error messages or if users would even notice a difference.

The Shneiderman guidelines state that error messages should say specifically what occurred (in language that the user can understand), be positive, and offer suggestions to the user on how to respond to the error (Shneiderman, 1998). Based on the Shneiderman guidelines, error messages with improved wording were developed, corresponding to the seven error messages identified in the previous section. The new error messages follow:

New Error Message 1:
"An error has occurred, which has forced the web browser application to close. This error may not have been caused by your actions. You might be able to operate without any problems by re-starting the browser application."

New Error Message 2:
"The web server that you requested is currently unavailable. This might be due to an incorrectly spelled web address, or this might be due to an error at the web server. Please check and make sure that the address is spelled correctly. If that does not solve the problem, it might be helpful to try accessing the same web server at a later time."

New Error Message 3:
"You cannot access the web page that you requested, because the owner of that web page has not made the document public. If you want to view that web page, you must contact the owner of the web page directly. Their contact information might appear on other portions of their web site."

New Error Message 4:
"To access this web page, you must provide a username and password, assigned by the web site's owners. If you have a valid username and password, please check

your spelling and try again. If you do not have a valid username and password, you will not be able to access the web page."

New Error Message 5:
"The file name that you requested is currently unavailable. This might be due to an incorrectly spelled web address, or this might be due to an error at the web server. Please check and make sure that the web address is spelled correctly. If that does not solve the problem, it might be helpful to try accessing the same file at a later time."

New Error Message 6:
"The web server is taking too long to respond. This is not caused by your actions. Please attempt to access the same web server at a later time."

New Error Message 7:
"There are problems with the domain name service on your network, which is out of your control. Please contact your local network manager or your Internet Service Provider."

These new error messages were then implemented using HTML and JavaScript. Error messages could be implemented by either placing the error message directly on a Web page or by placing the error message in a dialog box superimposed on the center of the screen. Error messages in dialog boxes superimposed on the screen tend to be the most effective because users cannot miss them or ignore them (Lazonder & Meij, 1995). Therefore, we decided to implement our error messages as dialog boxes superimposed on the screen. Using the JavaScript "alert" command, the user would receive an error message, presented as a dialog box, when they clicked on a link. This type of JavaScript dialog box most closely relates to error messages because the user hears a beep sound when the alert box appears, and the user cannot continue with their tasks until they have hit "OK" on the error message. See Fig. 9.2 for an example.

To test whether the newly designed error messages were better understood and more helpful, we prepared a survey for the experiment. We first paired the existing error messages and our newly designed new messages. The old and new error

FIG. 9.2. An improved error message, implemented through the JavaScript "alert" command.

messages were not labeled as such, which might bias the user's perception of the error message. Instead, the 14 different error messages (7 old and 7 new) were simply labeled 1 through 14. In other words, participants would not know which one was the old existing error message and which was the newly designed error message. For each message, four questions were asked to evaluate each message, related to the Shneiderman guidelines:

1. I found this error message to be positive.

   1   2   3   4   5   6   7   8   9
   Strongly   Strongly
   Disagree   Agree

2. I understood what occurred.

   1   2   3   4   5   6   7   8   9
   Strongly   Strongly
   Disagree   Agree

3. I understood how to respond to the error.

   1   2   3   4   5   6   7   8   9
   Strongly   Strongly
   Disagree   Agree

4. How should you respond to this error?

For the first three questions, a traditional Likert scale of 1 to 9 was used, with 9 as strongly agree and 1 as strongly disagree. In addition, an open-ended question was asked as the fourth question, for any unexpected or surprising thoughts from users. Participants were asked to answer each of the four questions for each of the 14 error messages.

Experimental Conditions

Once the error messages and survey had been developed, the experimental protocol was submitted to the university institutional review board for human subjects and approved. The experimental protocol methodology was pretested with one user to determine if there were any major problems with the experimental protocol, which there were not. A computer lab at Towson University was obtained for the study. This lab contains 25 Dell PL/350 GX1 Pentiums, each with a 17-in. color monitor. In two different sessions, a total of 34 undergraduate students at Towson University participated in the experiment. The students represented many different majors on campus. After being given instructions on the protocol for the experiment, subjects were given a URL in which they would find instructions, as well as a set of screens that they would go through. Participants were also given instructions on completing the survey instrument. Subjects were instructed that if their computer froze or stopped working, they should simply change to another computer (there were a number of extra computers available). Once the experiment began, the participants could not speak with each other or ask questions.

RESULTS

Data was collected from 34 participants who took part in the experiment. The data collected from the old error messages was compared with the data collected from the new error messages. The goal was to test whether the mean scores were significantly different between the old error messages and the new error messages. A t test was employed because it can assess whether the means of two groups are statistically different from each other. The t test is one of the most commonly used of all statistical tests. The specific t test used is the paired t test, because we are investigating a variable in two groups where there is a meaningful one-to-one correspondence between the data points in one group and those in the other. The data was analyzed using the STATA statistical software.

We conducted 21 t tests (3 questions per error message × 7 pairs of error messages). Out of the 21 t tests performed, 19 show statistically significant differences, indicating that the newly designed error message has a statistically different (in these cases, larger) score than the existing error message. The details of the results are as follows:

Error Message 1

Old Error Message:
"This program has performed an illegal operation and will be shut down. If the problem persists, contact the program vendor."

New Error Message:
"An error has occurred, which has forced the web browser application to close. This error may not have been caused by your actions. You might be able to operate without any problems by re-starting the browser application."

For t-test result for this pair of messages, see Table 9.1

1. For question 1: The average for the old error message is 2.823529, and for the new error message it is 5.5. The difference is 2.676471. The T-value is 6.3803, and the p value is 0.0000.
2. For question 2: The average for the old error message is 3.764706, and for the new error message it is 5.588235. The difference is 1.823529. The T-value is −3.6442, and the p value is 0.0005.
3. For question 3: The average for the old error message is 4.558824, and for the new error message it is 6.764706. The difference is 2.205882. The T-value is −4.1861, and the p value is 0.0001.

TABLE 9.1
Comparisons for Error Message 1

Question	Old Mean	New Mean	Difference	T-Value	p Value	Whether < 0.05
Question 1	2.823529	5.5	−2.676471	−6.3803	0.0000	Yes
Question 2	3.764706	5.588235	−1.823529	−3.6442	0.0005	Yes
Question 3	4.558824	6.764706	−2.205882	−4.1861	0.0001	Yes

TABLE 9.2
Comparisons for Error Message 2

Question	Old Mean	New Mean	Difference	T-Value	p Value	Whether < 0.05
Question 1	5.352941	6.735294	−1.382353	−2.6129	0.0067	Yes
Question 2	6.882353	8	−1.117647	−2.4819	0.0092	Yes
Question 3	7	8.294118	−1.294118	−2.7578	0.0047	Yes

Error Message 2

Old Error Message:
"Netscape is unable to locate the server: web.towson.edu. Please check the server name and try again."

New Error Message:
"The web server that you requested is currently unavailable. This might be due to an incorrectly spelled web address, or this might be due to an error at the web server. Please check and make sure that the address is spelled correctly. If that does not solve the problem, it might be helpful to try accessing the same web server at a later time."

For the *t*-test result for this pair of messages, see Table 9.2

1. For question 1: The average for the old error message is 5.352941, and for the new error message it is 6.735294. The difference is 1.382353. The T-value is 2.6129, and the p value is 0.0067.
2. For question 2: The average for the old error message is 6.882353, and for the new error message it is 8. The difference is 1.117647. The T-value is 2.4819, and the p value is 0.0092.
3. For question 3: The average for the old error message is 7, and for the new error message it is 8.294118. The difference is 1.294118. The T-value is 2.7578, and the p value is 0.0047.

Error Message 3

Old Error Message:
"Forbidden—You don't have permission to access www.towson.edu/~jlazar/forbidden.html on this server."

TABLE 9.3
Comparisons for Error Message 3

Question	Old Mean	New Mean	Difference	T-Value	p Value	Whether < 0.05
Question 1	4.147059	6.382353	−2.235294	−4.1461	0.0001	Yes
Question 2	6.735294	8.029412	−1.294118	−2.5765	0.0073	Yes
Question 3	6.294118	7.558824	−1.264706	−2.2005	0.0174	Yes

New Error Message:
"You cannot access the web page that you requested, because the owner of that web page has not made the document public. If you want to view that web page, you must contact the owner of the web page directly. Their contact information might appear on other portions of their web site."

For the *t*-test result for this pair of messages, see Table 9.3

1. For question 1: The average for the old error message is 4.147059, and for the new error message it is 6.382353. The difference is 2.235294. The T-value is 4.1461, and the p value is 0.0001.
2. For question 2: The average for the old error message is 6.735294, and for the new error message it is 8.029412. The difference is 1.294118. The T-value is 2.5768, and the p value is 0.0073.
3. For question 3: The average for the old error message is 6.294118, and for the new error message it is 7.558824. The difference is 1.264706. The T-value is 2.2005, and the p value is 0.00174.

Error Message 4

Old Error Message:
"Authorized failed. Retry?"

New Error Message:
"To access this web page, you must provide a username and password, assigned by the web site's owners. If you have a valid username and password, please check your spelling and try again. If you do not have a valid username and password, you will not be able to access the web page."

For the *t*-test result for this pair of messages, see Table 9.4

1. For question 1: The average for the old error message is 4.323529, and for the new error message it is 6.852941. The difference is 2.529412. The T-value is 5.2700, and the p value is 0.0000.
2. For question 2: The average for the old error message is 4.588235, and for the new error message it is 8.176471. The difference is 3.588235. The T-value is 6.4408, and the p value is 0.0000.

TABLE 9.4

Comparisons for Error Message 4

Question	Old Mean	New Mean	Difference	T-Value	p Value	Whether < 0.05
Question 1	4.323529	6.852941	−2.529412	−5.2700	0.0000	Yes
Question 2	4.588235	8.176471	−3.588235	−6.4408	0.0000	Yes
Question 3	5.676471	8.058824	−2.382353	−4.4889	0.0000	Yes

TABLE 9.5

Comparisons for Error Message 5

Question	Old Mean	New Mean	Difference	T-Value	p Value	Whether < 0.05
Question 1	5.117647	6.264706	−1.147059	−3.0541	0.0022	Yes
Question 2	5.617647	7.617647	−2	−4.1708	0.0001	Yes
Question 3	5.588235	7.852941	−2.264706	−4.4499	0.0000	Yes

3. For question 3: The average for the old error message is 5.676471, and for the new error message it is 8.058824. The difference is 2.382353. The T-value is 4.4889, and the p value is 0.0000.

Error Message 5

Old Error Message:
"The web server can not find the file or script you asked for. Please check the URL to ensure that the path is correct. Please contact the server's administrator if this problem persists."

New Error Message:
"The file name that you requested is currently unavailable. This might be due to an incorrectly spelled web address, or this might be due to an error at the web server. Please check and make sure that the web address is spelled correctly. If that does not solve the problem, it might be helpful to try accessing the same file at a later time."

For the t-test result for this pair of messages, see Table 9.5

1. For question 1: The average for the old error message is 5.117647, and for the new error message it is 6.264706. The difference is 1.147059. The T-value is 3.0541, and the p value is 0.0022.
2. For question 2: The average for the old error message is 5.617647, and for the new error message it is 7.617647. The difference is 2. The T-value is 4.1708, and the p value is 0.0001.
3. For question 3: The average for the old error message is 5.588235, and for the new error message it is 7.852941. The difference is 2.264706. The T-value is 4.4499, and the p value is 0.0000.

TABLE 9.6
Comparisons for Error Message 6

Question	Old Mean	New Mean	Difference	T-Value	p Value	Whether < 0.05
Question 1	5.235294	6.382353	−1.147059	−3.3159	0.0011	Yes
Question 2	7	7.382353	−0.3823529	−0.8036	0.2137	No
Question 3	6.382353	7.647059	−1.264706	−2.6653	0.0059	Yes

Error Message 6

Old Error Message:
"There was no response. The server could be down or is not responding. If you are unable to connect again later, contact the server's administrator."

New Error Message:
"The web server is taking too long to respond. This is not caused by your actions. Please attempt to access the same web server at a later time."

For the *t*-test result for this pair of messages, see Table 9.6

1. For question 1: The average for the old error message is 5.235294, and for the new error message it is 6.382353. The difference is 1.147059. The T-value is 3.3159, and the p value is 0.0011.
2. For question 2: The average for the old error message is 7, and for the new error message it is 7.382353. The difference is 0.3823529. The T-value is 0.8036, and the p value is 0.2137. There was not a statistically significant difference for error message 6, question 2.
3. For question 3: The average for the old error message is 6.382353, and for the new error message it is 7.647059. The difference is 1.264706. The T-value is 2.6653, and the p value is 0.0059.

Error Message 7

Old Error Message:
"Netscape is unable to locate the server. The server does not have a DNS entry. Check the server name in the location (URL) and try again."

New Error Message:
"There are problems with the domain name service on your network, which is out of your control. Please contact your local network manager or your Internet Service Provider."

For the *t*-test result for this pair of messages, see Table 9.7

1. For question 1: The average for the old error message is 4.676471, and for the new error message it is 5.617647. The difference is 0.9411765. The T-value is 1.9864, and the p value is 0.0277.

TABLE 9.7
Comparisons for Error Message 7

Question	Old Mean	New Mean	Difference	T-Value	p Value	Whether < 0.05
Question 1	4.676471	5.617647	−0.9411765	−1.9864	0.0277	Yes
Question 2	5.176471	6.058824	−0.8823529	−1.7077	0.0485	Yes
Question 3	5.382353	6.088235	−0.7058824	−1.1719	0.1248	No

2. For question 2: The average for the old error message is 5.176471, and for the new error message it is 6.058824. The difference is 0.8823529. The T-value is 1.7077, and the p value is 0.0458.

3. For question 3: The average for the old error message is 5.382353, and for the new error message it is 6.088235. The difference is 0.7058824. The T-value is 1.1719, and the p value is 0.1248. There was not a statistically significant difference for error message 7, question 3.

Data from the Open-ended Question

After viewing each error message, the subjects were asked to fill out three Likert scales on their perceptions of the error message. We felt that it was also important to ask the subjects a fourth question, an open-ended qualitative question on how they would respond to the error. The goal of this question was to determine if it was not only the participants' perception that they have a better idea of what occurred but also that they could more clearly understand and state what happened. If the subjects not only perceived that they had a better understanding but could also explain what happened, this would mean that they truly did have a better understanding of what occurred and what would be an appropriate response.

Although a complete list of responses to this question cannot be given in this chapter (14 error messages × 34 subjects would be very long), a sample of the responses will be given:

- For new error message 1, one participant indicated that they understood this "wasn't due to user action," which was unclear in the older error message. Another participant indicated with the new error message 1 that they should "try re-starting the browser."
- For new error message 2, one participant indicated that "if it doesn't work wait until later," whereas for the current error message 2, the participant indicated that they would just "try it again."
- For new error message 3, one participant indicated that "if I really need to access this page, I should contact the creator of the page."
- For new error message 4, one participant indicated that they should "enter user name and password," whereas with the current error message 4, they were

unsure as to what to do. With the current error message 4, another participant indicated they should "try again," but with the new error message 4 that they would need to "go get a username and password."

- For new error message 5, one participant indicated that they would check the URL and if "that doesn't work [they they would] try again later and it probably will."
- For new error message 6, one participant indicated they they should "just try again later," whereas with the current error message they indicated that "there is nothing you can do."
- For new error message 7, one participant indicated that they should "contact LAN manager or ISP," whereas with the current error message, the participant asked, "What is a DNS?" Another participant indicated that they would "wait until later, and then if it doesn't work, try to call your Internet Service Provider."

DISCUSSION

In 19 out of 21 responses comparing the current error messages to new error messages, the data analysis shows that subjects indicated statistically higher levels of comprehension and satisfaction in the newer error messages. The qualitative, open-ended responses also supported this analysis. In this experiment, the data can be interpreted to mean that users felt that the newer error messages were positive, better understood what occurred, and were more confident in responding to the error. An important question before the experiment was whether the users would notice a difference in error messages that were worded in a different manner. In this experiment, the users certainly did notice a difference in the error messages. This may mean that error messages with new wording could make a difference in how users perceive and respond to errors. This experiment helped to identify some common types of error messages that occur in the Web environment. In addition, possible approaches for future error message design were presented and were shown to be effective.

Research in the area of error messages is just beginning. There is a large amount of research work that still needs to be done in the area of user error. A number of steps should be taken in the future to further the research. First of all, a full taxonomy of user error must be developed. This taxonomy should take into account all causes of error, not only the situations that produce error messages, but also situations when users perceive that errors have occurred and errors that occur specifically in the networked environment. Another possible research direction is to take one specific error situation and to identify numerous possible error messages for it. Those error messages could then be tested to identify which error message was most effective.

SUMMARY

Good interface design assists the users in being productive and satisfied. Users are least productive and least satisfied when errors occur. Error messages must assist users in responding to the situation and continuing with their tasks. Many current error messages in software do not meet the most basic guidelines for a successful user experience. This experiment documented that error messages can be modified and improved to better meet the needs of users. With more comprehensive error messages, users can get a better understanding of their situation, respond to the error, and will feel more satisfied. More work must be done in the research area of error messages to enhance the user experience.

REFERENCES

Cable News Network. (2001). Test company: Error message didn't affect scores [Online, posted October 25, 2001]. Available: http://www.cnn.com
Carroll, J., & Carrithers, C. (1984). Training wheels in a user interface. *Communications of the ACM,* *27,* 800–806.
Hoffer, J., George, J., & Valacich, J. (2001). *Modern systems analysis and design* (3rd ed.). Reading, MA: Addison-Wesley.
Jacko, J. A., Salvendy, G., Sainfort, F., Emery, V. K., Akoumianakis, D., Duffy, V. D., Ellison, J., Gant, D. B., Gill, Z., Ji, G. Y., Jones, P. M., Karsh, B.-T., Karshmer, A. I., Lazar, J., Peacock, B., Resnick, M. L., Sears, A., Smith, M. J., Stephanidis, C., & Ziegler, J. (2001, in press). Intranets and organizational learning: A research and development agenda. *International Journal of Human–Computer Interaction, 14*(1), 95–130.
Lazar, J. (2001). *User-centered web development.* Sudbury, MA: Jones & Bartlett.
Lazar, J., & Norcio, A. (2000). System and training design for end-user error. In S. Clarke & B. Lehaney (Eds.), *Human centered methods in information systems: Current research and practice* (pp. 76–90). Hershey, PA: Idea Group Publishing.
Lazonder, A., & Meij, H. (1995). Error information in tutorial documentation: Supporting users' errors to facilitate initial skill learning. *International Journal of Human–Computer Studies, 42*(2), 185–206.
Nielsen, J. (2001). Error message guidelines [Online]. Available: http://www.useit.com/alertbox/20010624.html
Norman, D. (1983). Design rules based on analyses of human error. *Communications of the ACM, 26,* 254–258.
Norman, D. (1988). *The psychology of everyday things.* New York: HarperCollins.
Olsen, S. (2001, September 5). Microsoft gives error pages new direction. *The New York Times.*
Shneiderman, B. (1982). System message design: Guidelines and experimental results. In A. Badre & B. Shneiderman (Eds.), *Directions in human–computer interaction* (pp. 55–78). Norwood, NJ: Ablex.
Shneiderman, B. (1998). *Designing the user interface: Strategies for effective human–computer interaction* (3rd ed.). Reading, MA: Addison-Wesley.
Shneiderman, B. (2000). Universal usability: Pushing human–computer interaction research to empower every citizen. *Communications of the ACM, 43,* 84–91.

10

Character Grid: A Simple Repertory Grid Technique for Website Analysis and Evaluation "

by

Marc Hassenzahl (2003)

*Darmstadt University of Technology,
Institute of Psychology, Social Psychology
and Decision Making*

INTRODUCTION

People experience Websites as a whole instead of focusing on particular aspects. Its content, functionality, and interactional and visual style adds up to a holistic overall impression, a "character" (Janlert & Stolterman, 1997) or "product gestalt" (Monö, 1997). This character is a high-level description of a Website. It summarizes its numerous features, "behaviors," and appearance from a user perspective and, thus, represents the subjective overall perceptions, experiences, and evaluations of users. For example, in a study of online banking sites (Hassenzahl & Trautmann, 2001), participants experienced one site as very "analytical" and "professional." However, this assumed positive character was negatively evaluated. According to the participants, the design seemed to address experts rather than laypeople. They felt excluded. Another site was perceived as "playful," which in turn triggered worries about a probable lack of professionalism (i.e., "adequacy concerns," see also Hassenzahl & Wessler, 2000).

The function of assigning a character to a Website is to reduce cognitive complexity and, more importantly, to trigger particular strategies to handle the site. In the examples above, the perceived character of the Website led to far-reaching

183

inferences, which will eventuate in behavioral consequences, such as avoiding the site.

Given the assumed significance of the character of a Website, one might call for a method to explore and evaluate this character rather than isolated aspects (e.g., its usability). Because empirical evaluation is one of the cornerstones of user-centered design (e.g., Gould & Lewis, 1985; ISO, 1999), this method should be empirical and user driven—not analytical and expert driven.

PERSONAL CONSTRUCTS AND THE REPERTORY GRID TECHNIQUE

In his "Psychology of Personal Constructs," Kelly (1955) assumes that individuals create so-called personal constructs through which they filter all their experiences. An individual's unique collection of personal constructs, the personal construct system, determines the meaning attached to events or objects and thereby the way the individual approaches and interacts with the environment. They are personal "networks of meaning" (Fransella & Bannister, 1977); they define our subjective reality.

More specifically, personal constructs are conceptualized as dissimilarity dimensions. For example, I may characterize the difference between two Websites with the personal construct "fancy–conservative." On the one hand, my personal construct indicates something about the attributes I pay attention to or the topics that are relevant for me. From this, conclusions may be drawn about me as a person, for example, my experiences, values, and concerns. On the other hand, it also reveals information about the Websites (i.e., their attributes). From a design perspective, the information about differences between Websites (i.e., products) is more important than information about differences between individuals. Accordingly, the Character Grid will focus on the question of what personal constructs indicate about Websites.

The information embedded in the personal constructs is subjective in nature. On the person level, personal constructs can reflect the individual's experiences, values, and concerns. On a group level, they reveal which experiences, values, and concerns are commonly held (i.e., by different people) and which are ideographic (i.e., by only one individual). A better understanding of users' experiences is especially important in the context of the World Wide Web. Users can choose among a number of different companies offering similar content or goods. Competitors are "just one click away," and users tend to quickly leave sites on the basis of only cursory first impressions. Thus, the "subjective reality" of users when interacting with a Website is a valuable source of information.

The Repertory Grid Technique systematically extracts personal construct systems. It is a structured interviewing technique that typically consists of the two parts: personal construct elicitation and object rating. For construct elicitation, individuals are presented with a triad from a set of relevant objects (i.e., Websites)

TABLE 10.1
Part of a Personal Construct System (Participant R) and the Scores of Eight
Different Online Banking Sites

Negative (−)	Positive (+)	New Design	Old Design	Advance	Bawag	Commerz	Dresdner	OnVista	Sparda
neutral color −	provoking color	−	−	0	+	−	0	0	−
pictures of happy people − (home page)	information at the first glance (home page)	0	+	+	+	0	−	0	0
suitable for the mass market −	not suitable for everybody	0	+	0	0	−	−	0	−
many details −	clear visual presentation	−	+	+	+	0	0	−	+
pictures animated −	pictures not animated	+	0	+	+	0	0	0	0
obvious advertising −	trustworthy	0	0	+	+	+	0	−	+
sober −	promises fun	−	+	0	+	−	−	−	−

Note. A "+" indicates that the positive pole of the construct is applicable to the respective Website; a "−" indicates that the negative pole of the construct is applicable to the respective Website. A "0" indicates that neither the positive nor the negative pole of the construct is applicable to the respective Website. All constructs were originally in German.

to be compared. They are asked to indicate in what respect two of the three objects are similar to each other and differ from the third. This procedure leads to the generation of a personal construct that accounts for a difference. (A personal construct consists of two mutually exclusive poles, such as "playful–serious," "two-dimensional–three-dimensional," or "ugly–attractive".) In some instances, the individual indicates which of the two poles is perceived as desirable (i.e., has positive value). The result is an ideographic differential, comparable to a semantic differential (e.g., Osgood, Suci, & Tannenbaum, 1957). The left two columns of Table 10.1 show the personal construct system of a single participant from a study of eight different online banking sites (Note: All examples presented in this chapter are taken from a Repertory Grid study in the online banking domain, which is in part reported in Hassenzahl & Trautmann, 2001; Fig. 10.1 shows screen shots of the eight online banking sites).

In the second part, the object rating, the individuals are asked to rate all objects in the set on their individual personal constructs (on a scale anchored by the construct poles). The result is an individual-based rating of the objects based on differences between them. This is the basis for further analysis, such as computing similarities between Websites (later described). The eight columns on the right of Table 10.1 show the participant's rating of the eight Websites on his or her personal construct system. For example, the "bawag" site is the only site experienced as having a "provocative color." Note, however, that the personal constructs do not necessarily have to be explicitly positive or negative, as in the example given.

The Repertory Grid Technique is applied in various fields, ranging from clinical psychology (for a collection of clinical applications see Slater, 1976) to knowledge

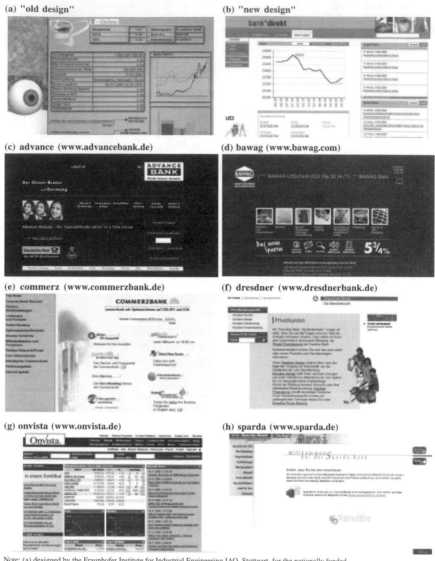

(a) "old design"

(b) "new design"

(c) advance (www.advancebank.de)

(d) bawag (www.bawag.com)

(e) commerz (www.commerzbank.de)

(f) dresdner (www.dresdnerbank.de)

(g) onvista (www.onvista.de)

(h) sparda (www.sparda.de)

Note: (a) designed by the Fraunhofer Institute for Industrial Engineering IAO, Stuttgart, for the nationally funded INVITE research project on Human-Computer interaction, (b) designed by User Interface Design GmbH, Munich, for INVITE, (c) Advance Bank AG, Munich (d) Bank für Arbeit und Wirtschaft AG, Vienna, Austria, (e) Commerzbank AG, Franfurt/Main, (f) Dresdner Bank AG, Frankfurt/Main, (g) OnVista AG, Cologne, and (h) Verband der Sparda-Banken e.V., Frankfurt/Main. All sites except (a) and (b) were used in the versions available in October and November 2000.

FIG. 10.1. The eight different online banking sites used as examples in the present chapter (taken from Hassenzahl & Trautmann, 2001).

acquisition (e.g., Gaines & Shaw, 1997). The results proved helpful for purposes as diverse as structuring hypertexts (Dillon & McKnight, 1990) or exploring textile qualities (Moody, Morgan, Dillon, Baber, & Wing, 2001).

Despite this, in the context of product evaluation in general or Website development, applications of the Repertory Grid Technique are rather rare. Several authors suggest that the Repertory Grid Technique can be useful in product evaluation (Baber, 1996; Sinclair, 1990) and may "significantly broaden usability engineering practices by shifting the focus to a more holistic perspective on human needs and desires" (Hassenzahl, Burmester, & Beu, 2001, p. 75). For example, a Repertory Grid study of prototypes for control room software (Hassenzahl & Wessler, 2000) showed that the gathered personal constructs addressed not only usability (i.e., quality of presentation and interaction) but also issues such as "adequacy concerns" (e.g., concerns about whether the prototype is suitable for its intended context of use) or "hedonic quality" (i.e., task-unrelated qualities, such as modernity or ability to stimulate; see Hassenzahl, Burmester, & Beu, 2001). Importantly, 36% of the elicited personal constructs were found to be highly design relevant, that is, useful for product redesign (e.g., the font size is too small or font size is appropriate). Thus, the basic idea of personal constructs and the Repertory Grid Technique, respectively, can be helpful for the evaluation of products in general and of Websites in particular.

REQUIREMENTS FOR A REPERTORY GRID TECHNIQUE SUPPORTING EMPIRICAL WEBSITE EVALUATION IN AN INDUSTRIAL CONTEXT

The Repertory Grid Technique is a method that comes in many variants. More than being a definite method, it has to be tailored to the special requirements of a given application domain. The purpose of the present chapter is to present a variant of the Repertory Grid Technique that will prove useful for the empirical exploration and evaluation of a Website's character in an industrial context. I call this variant the Character Grid. It can be understood as a character study of the respective Websites. To be useful, it has to fulfill the following three requirements: (1) focus on design-relevant qualitative data, (2) enable selection between "good" and "bad" design ideas, and (3) be simple.

Focus on Design-Oriented Qualitative Data

The primary goal of all the methods employed in the course of a design process is the emergence of a new system. Rich and detailed information is needed to help guide a continuous process of design improvement. Thus, the Character Grid will focus on the content of the personal constructs (e.g., which general and ideographic topics are addressed) rather than on the structure of the personal construct systems (e.g., which constructs belong together). How these data are design oriented is determined by the way in which they are analyzed (described later).

Enable Selection Between "Good" and "Bad" Design Ideas

The design process is constant problem solving and decision making. At the beginning, there seems to be a boundless number of design possibilities, but in the course of the design, this number is eventually narrowed down until a final design is reached. This is accomplished by separating "good" design ideas (i.e., positively valued personal construct poles) from "bad" design ideas (i.e., negatively valued personal construct poles). Therefore, constructs that express no values are of limited benefit for guiding design. For example, in online banking, it is not enough to just know that advertising is a topic for users as reflected by personal constructs such as "with advertising–no advertising" (participant B12) or "obvious advertising–trustworthy" (participant R). The problem is that the first construct leaves it to our interpretation whether the participant liked the idea of having advertising on the site or not because his embedded values are implicit. In the second construct, the value is more explicit. "Trustworthy" is surely a desired attribute of a banking site, and the "obvious advertising" seems to corrupt a Website's trustworthiness–at least in the eyes of the participant.

In the Character Grid, measures are taken to reduce the number of merely descriptive constructs. The requirement "enable selection" is satisfied in the construct elicitation part of the Character Grid (defined later).

Be Simple

The industrial context calls for simplified methods because a method's complexity can be a hindrance to employ in the daily routine (Bellotti, 1988). Methodological complexity can be problematic in at least two ways. First, an in-depth knowledge is necessary or expected that is often simply not available. The field of human–computer interaction is a wide and strongly multidisciplinary field. It is difficult, if not impossible, for practitioners of the field to keep track of both technological and methodological progress and their daily work. Second, something like a "silver bullet" method—a secret weapon that gathers all information needed in one go—does not exist. A combination of methods is used in an industrial context (e.g., usability testing with additional interviews or questionnaires), leaving each method with a smaller share of time and money.

Practical Website design (and evaluation) is not rocket science. The methods used should be simple and easy to apply. They are tools for gathering and catalyzing insights rather than for discovering an "ultimate truth," as requested by most scientific methods. The requirement "be simple" is fulfilled in the way the obtained data are analyzed (covered later). The Character Grid, described in the following section, is aimed at fulfilling the three requirements necessary for a method to be useful in an industrial context.

THE CHARACTER GRID: PLANNING

Before a Character Grid study can be carried out, the study participants and the Websites included in the study have to be carefully selected.

Participants

A Character Grid study can be carried out with two different goals in mind. On the one hand, you may want to focus on certain target user groups, such as "stock market experts" or "banking laypeople." In this case, a careful selection of participants based on a screening procedure during recruiting is necessary. If your recruiting is successful, you will have a "homogenous" sample and will find a strong overlap in the participants' perspectives (i.e., personal construct systems). This overlap makes it easy to get a coherent understanding of how the various Websites are experienced and evaluated by one of the target user groups.

On the other hand, you may want to collect as many different perspectives (i.e., personal construct systems) about your Website as possible. This strategy is useful if the target user group is quite diverse or if you want to be sure not to miss any possible topic. In this case, you should try to maximize the differences between the participants. Try to vary as many dimensions (e.g., age, internet expertise, attitude toward product, etc.) within your sample as possible. Such a "heterogeneous" sample will result in less overlap in the collected personal construct systems. In contrast to the results obtained with the homogenous sample, individuals' ideographic views will be more prominent. This lack of overlap makes it harder to get a coherent understanding of how the Websites included in the study are experienced and evaluated. However, especially in early phases of the Website design process, the diversity in perspectives can foster a better understanding of the topics likely to arise in later stages of the design process and may stimulate designers.

Regardless of whether your study goals call for a homogenous or a heterogeneous sample, because of the Character Grid's focus on ideographic views, it is necessary to know something about the "background" of each participant. Try to get to know each participant—her or his general view of things—to be able to better understand differences in the obtained personal construct systems. You must be aware that in contrast to more quantitatively oriented methods, the Character Grid deeply probes into the experiences and evaluations of individuals.

To generalize from the obtained, ideographic personal construct systems and to be able to explain deviations from expected constructs systems requires background knowledge about the various individuals that hold these construct systems. Most of the time, a "debriefing" interview after the study is the right setting in which to learn a bit more about the participants, their background, and general view of things (addressed later).

A last important question concerns the number of participants for a study. In an industrial context, this is a question of resources rather than of scientific scrutiny. All analysis techniques proposed in the section "The Character Grid: Analysis" are working with only a single participant. However, you must be aware that a study with an individual is no more than the reconstruction of this individual's ideographic perspective. A single-person Character Grid study can be helpful when done with the lead designer of the Website to be designed. It may systematically reveal the experiences and evaluations the designer seeks to create and thus will foster mutual understanding and communication in the design team. Moreover, the personal construct system of the designer (or designers) may serve as a baseline for later studies with potential users. For example, the designer's general idea of the desired Website's character might be appropriate, but he or she may have not succeeded in properly communicating this character to potential users. A comparison of the personal construct systems of the designer and potential users can clarify such situations.

From my experience, the ideal number of participants for a Character Grid study for obtaining representative data while sparing resources is 10. With a homogeneous sample of 10 participants you can expect an overlap of personal construct systems and thus the emergence of shared topic (covered later). With a heterogeneous sample of 10, you can expect a wide variety of different personal construct systems and thus the desired diversity in perspectives.

Websites

The Websites included in the study should be chosen in a way so that they represent the "design space" to be analyzed. The "design space" (see Moran & Carroll, 1994, for an overview) is the "bundle" of design-driving information, such as the general purpose of the Website, its context of use (Bevan & Macleod, 1994), and connected trade-offs and arguments. In the context of Website design, the studied Websites will most likely consist of a "new design," the succeeding design, and major competitors in the market. Thus, the selection of Websites marking the design space is more or less predetermined. In general, I found a number of five to eight Websites in one study convenient.

Sometimes, a large number of competitor Websites are available or, in the case of new business ideas, competitors are hard to determine. In these cases, the selection of Websites should be based on the following two rules: (1) choose representative Websites and (2) maximize the diversity (i.e., heterogeneity) among the representative Websites.

To choose representative Websites requires the careful consideration of the scope of the study. A good heuristic for determining whether Websites are representative for your scope or not is to ask yourself how enlightening a comparison between these Websites and your "new design" would be. If, for example, you are interested in whether your Website can compete with the best Websites known, you could include some of the "most popular Websites" along with your new design in your study.

Even if you have properly identified the scope of the study and a possible set of representative Websites, you may have to further narrow down the set. Here, the second rule should be applied, namely to maximize the diversity (i.e., heterogeneity) among the representative Websites in the set. One of the core assumptions of the Repertory Grid is that a personal construct is equal to a dissimilarity dimension triggered by variation between Websites. In other words, attributes of Websites that may be important but do not vary in the set of studied Websites will never surface as a personal construct—they will go unnoticed. For example, personal constructs like "suitable for the mass market—not suitable for everybody" are unlikely to emerge without having a Website in the set that deviates from well-known design styles. This emphasizes the importance of maximizing the heterogeneity of the Websites selected for the study. Thus, even "deviating" Websites that seem to comprise extreme design means are desirable as a set member. The more differences that exist and the more extreme they are, the more unlikely it becomes that relevant topics will go unnoticed.

THE CHARACTER GRID: PROCEDURE

A single Character Grid interview will take 2 to 3 hours. It consists of four parts: (1) introduction, (2) construct elicitation, (3) Website rating, and (4) debriefing.

Introduction

The main goal of the introduction is to familiarize the participants with each Website. The simplest form is to just show some screen shots of each Website to the participant. The main drawback of this approach is the resulting strong focus on appearance, which is likely to be reflected in personal constructs that revolve around differences in "look" (e.g., "too much information–appropriate amount of information," "matter-of-factly–warm colors") rather than "feel." To increase the likelihood of personal constructs that point at differences in the interaction with the Websites (i.e., "feel"), participants should be given 2 to 3 minutes to "play around" with each Website. This interaction time can be further standardized by providing tasks for each Website (quite similar to a task in a usability test). However, depending on the number of Websites in the set, the latter approach may be too time-consuming.

Construct Elicitation

For construct elicitation, the participant is presented with a randomly drawn triad from the set of Website designs to be compared. She or he is asked to "think of the following three Websites in the context of designing a new Website with a similar general purpose. In what way are two of them alike and different from the third?" First, the participant is asked to select the odd one out and to give a term (i.e., a construct pole) "that characterizes the way in which the selected one is different." Second, a term (i.e., a construct pole) has to be given that "characterizes the way in which the other two are alike." This results in a personal construct that accounts for a perceived difference, such as "playful–serious," "ugly–attractive." Sometimes, an additional discussion about the construct poles provided is necessary, which may result in a rephrasing (described later).

After the construct is clearly stated, the participant indicates which of the two construct poles is perceived as desirable (i.e., has positive value). The process is repeated until 15 constructs are elicited. After the elicitation is finished, you should plan a small break. There are several points to consider while running the construct elicitation procedure:

• **Presentation of the Websites in the triad.** To elicit valid constructs, it is crucial for the participants to remember attributes of the Websites in the triad from the introduction. To facilitate the recollection, you may provide screen shots or "screencam" movies (e.g., "HyperCam," http://www.hyperionics.com/hc/index.html) of the Websites in the triad.

• **Labeling of construct poles.** You must avoid superficial and vague constructs, such as "good–bad" or "nice–ugly." If a participant produces such a construct you should try to further explicate what exactly is "bad" or "ugly" and what is "good" or "nice." You should be aware that the actual process of labeling and rephrasing of the construct poles (i.e., finding a describing term) requires verbal skills in both the experimenter and the participant. It can be best described as "mining" for the core meaning of the participant's experienced difference between the Websites.

• **Novel constructs.** Each newly elicited construct should substantially differ from the constructs elicited so far. Whenever a newly elicited construct is very close to one of the constructs already elicited, you should request a further explication of the differences from this older construct. This explication process may yield rephrased construct poles, which makes the construct more comprehensible and different from the other constructs. If this doesn't work, you should simply skip this triad of Websites and create a new one.

• **Clear preferences.** A purely descriptive construct (e.g., "two-dimensional–three-dimensional") is of limited value when it comes to improving a Website. Probes into people's preferences are needed (e.g., to distinguish the "good" from the "evil"). Thus, all constructs for which the participants have difficulties in stating clear preferences (i.e., which pole is positive and which pole is negative) should be rejected.

• **Documentation.** Use a form similar to Table 10.1 to document the elicited personal constructs (after all necessary clarifications or rephrasing).

Website Rating

In the Website rating, the participant has to rate all Websites on her or his personal constructs. If you use a form similar to Table 10.1, you may instruct the participant to put in a "+" if the positive pole of the construct applies to the Website, a "−" if the negative pole applies, and a "0" if neither the positive nor the negative pole applies. The result is an individual-based description of the Websites based on differences among them.

Debriefing Interview

In the debriefing interview, the interviewer should try to shed light on the participants' background and their general view of things. Here, a combination of an open, unstructured interview and some questionnaires (e.g., to document demographics, to formally assess computer expertise, etc.) is appropriate. I strongly recommend that the additional questioning be done at the end of the Character Grid interview. Otherwise, it becomes likely that the labeling of the constructs, and thus the mining for the core meaning, is colored by your preconceptions about the participant. Further, you should avoid seeking only information in favor of "hypotheses" (or opinions) you have generated during the earlier parts of the Character Grid interview. In contrast, you should explicitly look for contradictions that may correct any unjustified opinions you hold about the participant and her or his perspective. To give an example: A participant in the online banking study may generate constructs that value "playfulness" and "to be different" as positive and value typically positive attributes associated with banking sites, such as "serious" or "matter-of-factly," as negative. As an interviewer, you may tend to view this participant as dishonest or provoking. Based on one of these hypotheses, you will start to discount her or his perspective on the Websites. You may convince yourself that the constructs stated are not the participant's "true" opinions, that he has a "hidden agenda." In the debriefing interview, you must take the chance to falsify this hypothesis rather than verify it. You may tell the participant that his preference is the opposite of what other participants prefer or what you as an interviewer expected. Ask him to find an explanation for this. You may ask about her or his preferences in other domains (e.g., governmental sites) to get an idea of how wide ranging her or his preference for "playfulness" is. To summarize, give the participant a chance to clarify her or his perspective, and don't readily jump to unjustified conclusions.

THE CHARACTER GRID: ANALYSIS

The analysis techniques presented in this section satisfy the requirements formulated in a previous section. They focus on qualitative data, aim at distinguishing between "good" and "bad," and are simple. Admittedly, they are rather crude heuristics that can be definitely refined. However, in my experience, they are helpful for understanding users' perceptions, experiences, and evaluations.

Appeal: Which Website Is the Best, Which Is the Worst?

In the beginning, it should be determined which of the studied Websites was experienced as appealing and which was not. On the basis of the ratings of each Website on each personal construct, a ranking of appeal can be computed easily. First, you count all positive ratings per Website (indicated by a "+"). From this number, subtract the number of negative ratings per Website (indicated by a "−"). Rank order the Websites according to the resulting index. The higher the index, the better the site. If two sites share the same index (i.e., a "tie"), they are assigned the same rank.

Table 10.2 shows an "appealingness ranking" for a single participant (participant R) as an example. (Note: To facilitate understanding of the applied analysis technique, I will use only data from a single person in the present and following section. However, the same analysis can be applied to the whole data set.) It can be seen that the "bawag" design was experienced as the most appealing and the "OnVista" design as the most unappealing. "Commerz" and "new design" share the same index. They are assigned the same rank (a "tied" rank of 6).

You may ask yourself whether the proposed method is a valid strategy. To explore this, we made an additional step in the online banking site study (Hassenzahl & Trautmann, 2001), which is not a part of the Character Grid. In between construct elicitation and Website rating, we asked people to rank order the Websites according to their appeal. Based on these individual rankings, an "overall ranking" can be computed by averaging the individual ranks. The correlation between the overall ranking and the ranking derived on the basis of the "number of positive ratings minus number of negative ratings" heuristic is .92 (Spearman's ρ, $p < .000$, $N = 8$, two-tailed). This correlation is pretty high (1 would be the

TABLE 10.2

An "Appealingness Ranking" On the Basis of Participant R's Construct System
and His Ratings

Negative (−)	Positive (+)	New Design	Old Design	Advance	Bawag	Commerz	Dresdner	OnVista	Sparda
neutral color –	provoking color	−	−	0	+	−	0	0	−
pictures of happy people – (home page)	information at the first glance (home page)	0	+	+	+	0	−	0	0
poor –	concise	0	+	+	+	+	+	0	0
evaluative pictures –	illustrative pictures	0	0	0	0	0	−	0	0
conservative presentation –	humorous presentation	−	+	0	0	−	−	0	0
too much information –	clear structure	−	+	+	+	0	0	−	+
serious –	makes curious	0	+	0	+	−	−	−	0
no news on home page –	news on home page	0	+	+	0	0	0	0	0
adjusted –	shows courage to offend	−	+	0	+	−	−	0	−
for elderly people –	modern	0	+	0	+	−	0	+	0
suitable for the mass market –	not suitable for everybody	0	+	0	0	−	−	0	−
many details –	clear visual presentation	−	+	+	+	0	0	−	+
pictures animated –	pictures not animated	+	0	+	+	0	0	0	0
obvious advertising –	trustworthy	0	0	+	+	+	0	−	+
sober –	promises fun	−	+	0	+	−	−	−	−
Number of "+"		1	11	7	11	2	1	1	3
Number of "−"		6	1	0	0	7	7	5	4
Number of "+" – number of "−"		−5	10	7	11	−5	−6	−4	−1
Appealingness rank		6	2	3	1	6	8	5	4
					Best		*Worst*		

Note. A "+" indicates that the positive pole of the construct is applicable to the respective Website; a "−" indicates that the negative pole of the construct is applicable to the respective Website. A "0" indicates that neither the positive nor the negative pole of the construct is applicable to the respective Website. All constructs were originally in German.

maximum), which makes the proposed "appealingness ranking" a valid predictor of true subjectively experienced appeal.

Similarity: Which Websites Are Alike, Which Differ?

You may also be interested in the similarity (or dissimilarity) of the Websites in the set. For example, assume you are studying a new design that aims at differentiating the owning company from its stiffest competitor. To calculate the similarity between two Websites, count the number of constructs with identical ratings.

TABLE 10.3

Similarity Between Two Websites on the Basis of Participant R's Construct System and His Ratings

Negative (−)	Positive (+)	New Design	Advance	Same Rating?
neutral color −	provoking color	−	0	no
pictures of happy people − (home page)	information at the first glance (home page)	0	+	no
poor −	concise	0	+	no
evaluative pictures −	illustrative pictures	0	0	yes
conservative presentation −	humorous presentation	−	0	no
too much information −	clear structure	−	+	no
serious −	makes curious	0	0	yes
no news on home page −	news on home page	0	+	no
adjusted −	shows courage to offend	−	0	no
for elderly people −	modern	0	0	yes
suitable for the mass market −	not suitable for everybody	0	0	yes
many details −	clear visual presentation	−	+	no
pictures animated −	pictures not animated	+	+	yes
obvious advertising −	trustworthy	0	+	no
sober −	promises fun	−	0	no
Number of "yeses"				5
(Number of "yeses"/number of constructs) * 100				33%
Appealingness rank		6	3	

Note. A "+" indicates that the positive pole of the construct is applicable to the respective Website; a "−" indicates that the negative pole of the construct is applicable to the respective Website. A "0" indicates that neither the positive nor the negative pole of the construct is applicable to the respective Website. All constructs were originally in German.

Table 10.3 shows participant R's ratings of the new design and the "advance" site (as the assumed competitor). In the column "the same?" a "yes" is noted when the ratings on a construct are the same and a "no" when the ratings are different. For example, R experiences both sites as not having animated pictures—a fact he prefers. The amount of agreement in rating (i.e., number of "yeses") is the measure of similarity. It can be further standardized by dividing the result by the total number of constructs and multiplying this by 100. In our example, the number of "yeses" is 5, and the total number of constructs is 15. The similarity, thus, equals 33%. Therefore, the new design differs from the competitor. However, combined with the "appealingness rank" (see the previous section), it becomes apparent that in R's eyes the new design is less appealing. In other words, it differentiates, but it is not the usually desired "positive" differentiation.

Sometimes, it can be helpful to compute the similarities between all Websites. This is done exactly the way described above. Table 10.4 shows the matrix of

TABLE 10.4

Similarity Between All Web Sites on the Basis of Participant R's Construct
System and His Ratings

	New Design	Old Design	Advance	Bawag	Commerz	Dresdner	OnVista	Sparda
New design	100	20	33	27	47	40	53	60
Old design		100	40	60	27	20	20	33
Advance			100	60	20	20	33	47
Bawag				100	27	13	33	60
Commerz					100	67	40	60
Dresdner						100	33	40
OnVista							100	47
Sparda								100

Note. 100 indicates maximal similarity; 0 indicates no similarity.

all Websites and the according similarities. For example, the new design is most
similar to "sparda" and least similar to the "old design."

Comparison: Understanding the Differences

So far, we only answered questions about the differences in appeal or about the general similarities (or dissimilarities) of Websites. Design for improvement, however, requires a deeper understanding of these differences.

In practice, a commercial Website is often a redesign of an old site that is challenged by various competitors. Comparing these different designs, the following three questions are likely to arise:

- How does the character of the most appealing site differ from that of the least appealing site? This comparison reveals essential attributes leading to the perception of a top-quality design in the given domain.
- Does the new design communicate the intended character? How does it differ from the most appealing site? This comparison points at opportunities for the further improvement of a site.
- How do the characteristics of the new design and the old design differ (i.e., the predecessor design)? This comparison points to the progress (or regress) made with the new design.

Thus, three comparisons are typical: a *progress comparison* (i.e., new design vs. an old design), an *improvement comparison* (i.e., new design vs. the most appealing), and a *domain comparison* (i.e., most appealing vs. the least appealing).

TABLE 10.5

The Constructs of Participant R's Construct
System That Differentiate Between the "New Design"
and the Most Appealing Website (i.e., "Bawag")

	New Design	Bawag (Best)
1	too much information	– *clear structure*
2	many details	– *clear visual presentation*
3	sober	– *promises fun*
4	neutral color	– *provoking color*
5	adjusted	– *shows courage to offend*
6	for elderly people	– *modern*

Note. Construct poles printed in italics have positive
value ("+").

However, other comparisons can be fruitful, too, such as comparing the new design with a particular competitor.

For a comparison of two Websites, all constructs are selected that differentiate between both sites. A construct is defined as differentiating if one design is rated as positive ("+") while the other site is rated as negative ("−") on the *same* construct (and vice versa).

Table 10.5 shows all constructs of participant R's construct system for an improvement comparison (i.e., "new design" vs. the most appealing site "bawag").

Constructs 1 and 2 point to the impression that the amount of information given by the new design on a single screen is not appropriate. R seems to be overwhelmed by information. The new design is characterized as "sober" (construct 3), "neutral" (construct 4), and "adjusted" (construct 5)—all attributes seemingly appropriate for an online banking site. Surprisingly, R does not positively value these characteristics. He or she expects a Website to be distinct, that is, to be "provoking" (construct 4), to "show courage" (construct 5), and even to be "fun" (construct 3), rather than well behaved. This suggests that use of online banking Websites can be construed as a "consumption experience" rather than purely utilitarian. With the last construct (6), R acknowledges that his preferences may differ from that of other user groups. He prefers an online banking Website that is "modern" (construct 6) and not "for elderly people." One cannot ignore the somewhat mocking connotation of the construct; however, it demonstrates that R is aware of his, to some extent, divergent needs (i.e., online banking as a consumption experience).

The analysis of the differentiating constructs as done above reveals interesting information. It shows that the "new design" is experienced with reduced usability due to information overload. In addition, it is not experienced as "hedonic," that is, it lacks task-unrelated quality that addresses the human need for novelty and change (i.e., stimulation) and the need for expressing oneself through objects (i.e., modernity; see Hassenzahl, Burmester, & Beu, 2001; Logan, Augaitis, & Renk, 1994).

Comparison: Identifying Shared Topics and Differing Experiences and Evaluations

It is important to keep in mind that the preceding example was based on a single individual. To get a true understanding of the differences between two Websites, the differentiating constructs of all participants have to be taken into account.

To define a topic, similar constructs have to be sorted into groups. A number of homogeneous groups (i.e., a topic system) will emerge. Some constructs will be difficult to assign to a certain group. These should be collected in a "miscellaneous" group. A good topic system has a small number of constructs in the "miscellaneous" group. Moreover, there shouldn't be too many constructs that you want to assign to more than one topic. If you encounter problems with overlapping topics, try to redefine the topics.

It is always helpful to ask a colleague to review the topic system you set up. You can even ask a colleague to come up with his own proposal for a topic system. The discussion about the differences between these topic systems will give you insights that will inevitably lead to a better topic system.

Table 10.6 shows all constructs that differentiate between the new design and the most appealing Website (i.e., "bawag"), grouped by topic.

For each topic, a "shared topic" index summarizes how many different participants contributed at least one construct to the topic. This number can be further standardized by dividing it by the number of participants and multiplying the result by 100. For example, the five constructs forming the topic "clarity of presentation" were contributed by 4 different participants (i.e., K, L, R, and W; see Table 10.6). Altogether, 10 individuals participated in the study. This equals a "shared topic" index of 40% ([4/10] * 100). If each participant contributes at least one construct to a topic, the index equals 100%—the topic is of general concern among the participants who experienced differences among the compared Websites. If only one participant contributes to a topic, the topic is purely ideographic and the index meaningless. Topics of general concern should be viewed as more important than ideographic topics.

A second interesting source of information is dissent in the participants' experiences and evaluations. For example, in the topic "news oriented," the same construct "no news ticker–news ticker" (constructs 23 and 24) was contributed by different participants (see Table 10.6). Interestingly, for participant W, the news ticker had positive value, whereas it had negative value for participant P. Such a discrepancy puts issues into focus that need additional careful consideration in the next design iteration. Also, however, a topic itself can cause dissenting experiences and evaluations. For example, consider the topic "stimulation," in Table 10.6. The same background color can be experienced as either negatively "aggressive" (construct 9) and "obtrusive" (construct 6) or positively "provoking" (construct 7). Such a discrepancy illuminates the consequences of the two alternatives—to be provoking at the risk of being perceived as aggressive or to be

TABLE 10.6
All Constructs That Differentiate Between the "New Design" and the Most
Appealing Web Site (i.e., "Bawag") Grouped by Topic

Participant		New Design	Bawag (Best)	Shared Topic Index
		Topic: clarity of presentation		
1	K	terse –	"generous" screen layout	4/10
2	L	many details –	clear	(40%)
3	R	too much information –	clear structure	
4	R	many details –	clear visual presentation	
5	W	frugal –	too much information on one page	
		Topic: stimulation		
6	P	calm color scheme –	obtrusive color scheme	3/10
7	R	neutral color –	provoking color	(30%)
8	W	color secondary –	color is important/in the fore	
9	W	serious color –	aggressive color	
		Topic: target group/purpose		
10	B1	specialized –	general purpose	4/10
11	F	focuses on stock market –	wide variety of different products	(40%)
12	K	target user group is vague –	target user is obvious	
13	L	for businesspeople –	for laypeople	
		Topic: rationality/emotionality		
14	B1	sober –	playful	4/10
15	K	matter-of-fact –	warm colors	(40%)
16	L	scientific –	information directly accessible	
17	L	analytical –	customer friendly	
18	R	adjusted –	courage to offend	
19	R	sober –	promises fun	
		Topic: news oriented		
20	B1	news oriented –	application-oriented	4/10
21	F	no choice concerning presentation of news –	news is optional	(40%)
22	F	current news on home page –	no news on home page	
23	P	no news ticker –	news ticker	
24	W	no news ticker –	news ticker	
		Topic: professionalism		
25	P	high-quality design –	nonprofessional design	2/10
26	W	professional –	nonprofessional	(20%)
		Miscellaneous		
27	W	two languages available –	one language only	
28	B3	without advertising –	with advertising	
29	P	search field available –	search field not available	
30	P	adaptable content –	fixed content	
31	P	fixed position for menu –	changing position for menu	
32	W	superfluous photos –	graphical	

Note. Construct poles printed in italics have positive value ("+"); topic labels are printed in bold.

serious at the risk to be too "neutral" and boring. This leads to informed design decisions.

Improvement: Turning the Data Into Potential "Change Actions"

To improve a Website design, decisions are required to mitigate the experienced negative aspects. To accomplish this, one may take a look at all the negative experiences people had with the site to be improved. A content analysis of the *improvement comparison*, that is, the site to be improved compared with the most appealing site (see the previous section), will show potential opportunities for improvement.

The results from the analysis described in the previous section (see Table 10.6) are the starting point for the analysis of improvement potential. The constructs per topic with their negative pole attached to the Website to be improved are summarized as *change actions*. A change action is a statement that describes the design goal of the action, an object to change, and the action itself.

Take, for example, the negative experiences of users of the "new design" Website grouped in the "clarity of presentation" topic (see Table 10.6, column "new design," constructs 1–4): "terse–generous screen layout," "many details–clear," "too much information–clear structure," "many details–clear visual presentation." The according change action is

To increase clarity of presentation	reduce	amount of visual detail/information
(Design goal)	(Action)	(Object)

In many cases, the design goal can be directly derived from the label of the topic, for example, "clarity of presentation." The constructs within the topic indicate objects, such as "information," "visual detail," "visual presentation," "screen layout," and "structure." In the example, the main object seems to be the amount of visual detail and information. The action itself is implied by construct content, such as "terse," "too much," and "many," namely to reduce the amount of visual detail and information.

Table 10.7 shows all derived change actions for the "new design" Website. The change actions vary in specificity. For example, change action 6 states that "to be more attractive include a news ticker." It is very specific—it states a clearly defined action and its object. By contrast, change action 3 lacks specificity. The analysis of the negative experiences expressed by the topic "target group/purpose" shows that somehow participants felt excluded by the apparent focus on the stock market and businesspeople. However, they are not very specific on how to change this impression by design.

A change action should be regarded as a "suggestion," a potential way to improve a Website. Obviously, each change action has to be evaluated before it is put into

TABLE 10.7

Derived Change Actions for the "New Design" Based on All Negative Constructs
Poles Attached to the Website*

Design Goal	Action	Object	Construct No.	Number of Constructs		
				Pro	Anti	Index
1. To increase clarity of presentation	Reduce	The amount of visual detail/information	1–4	4	1	3
2. To increase stimulation	Increase	The distinctiveness of the color scheme	7, 8	2	2	0
3. To appeal to laymen	Change	General impression	11–13	3	1	2
4. To be more emotionally appealing	Change	General impression	15, 16, 18, 19	4	2	2
5. To be more user friendly	Enable users to choose whether to show or not show	News	21	1	0	1
6. To be more attractive	Include	A news ticker	24	1	1	0
7. To be more attractive	Remove	Superfluous photos (on home page)	32	1	0	1

*See text for further details.

action. There are three criteria for evaluating change actions: (1) the desirability of the design goal, (2) the number of prochange constructs, and (3) the number of prochange versus antichange constructs.

The first criterion is the desirability of the change action's goal. Before changing the design according to the action, all involved stakeholders (i.e., product managers, marketing experts, designers, etc.) have to agree that the goal of a respective change action is a desired one. For example, the goal "to appeal to laypeople" can be contrary to an original design goal, such as "appeal to a stock market specialist." If so, the goal expressed by the participants may not be a desired one, and the stakeholders must make the deliberate decision to ignore the change action. The advantage of these discussions is twofold. First, all goals, their desirability and nondesirability, are explicitly stated and discussed. Thus, conflicting goals become obvious. Second, all design decisions are made deliberately and openly. Due to this, the acceptance of the design decisions among stakeholders is likely to increase, whereas the danger that the design follows "hidden agendas,"—unstated highly personal goals of single stakeholders—decreases. I believe that the discussion and negotiation processes fueled by the empirically derived change actions can have a positive impact on the quality of the final Website.

The second criterion is the number of prochange constructs, that is, constructs that support the change action. For example, the first change action in Table 10.7 is based on four prochange constructs, whereas the last change action is only based on one prochange construct. The higher the number of prochange constructs, the more seriously the change action should be taken. A low number of prochange constructs does not necessarily imply that the change action is worthless; it only points at the fact that the action (and its object) or the goal proposed by the change action may be ideographic. However, if the change action seems reasonable, one may put it into action nevertheless.

The third criterion is the prochange versus antichange constructs. If we take a closer look at the "clarity of presentation" topic (see Table 10.6, column "new design," constructs 1–5), we see that participant W experienced the new design oppositely. Instead of the Website being too detailed, she experienced it as "frugal," which had positive value to her. This is a construct that questions the change action and is therefore called an antichange construct. The number of antichange constructs is defined as the number of constructs per topic with their positive pole attached to the Website to be improved. For example, change action 2, "to increase stimulation, increase the distinctiveness of the color scheme," summarizes the two prochange constructs 7 and 8 of the "stimulation" topic ("neutral color," "color secondary"). However, the "stimulation" topic consists of two additional, antichange constructs, 6 and 9 ("calm color scheme," "serious color"), which address the same underlying idea but have their positive poles attached to the "new design" Website. Subtracting the number of antichange constructs from the number of prochange constructs results in a useful index for the evaluation of the change action:

• If the resulting number is positive, there is some agreement among participants that something is wrong with the site. The according change action is supported.

• If the resulting number is 0 (or close to 0), agreement is low among participants with regard to the positive or negative value of their experiences. Some participants experience, for example, the color scheme as negatively "neutral," whereas others experience the same color scheme as positively "calm" or "serious." The according change action needs careful consideration and additional analysis.

• If the resulting number is negative, there is some agreement among participants that the change recommended by the change action is not necessary. For example, if you would analyze the potential for improvement of the "bawag" instead of the "new design" Website according to the procedure described in this section, you would have created the change action "to improve clarity of presentation reduce information on one page" solely based on construct 5 (see Table 10.6, column "bawag [best] design"). This change action has one supporting construct and four dissenting constructs. The resulting index is negative (i.e., -3). The according change action is not supported and should not be put into action.

To summarize and discuss the potential for improvement is an important step. This is the point where the obtained data proves to be of direct, practical value.

For the sake of brevity, I proposed to take the *improvement comparison*—the comparison of the Website to be improved with the most appealing—as a starting point for the analysis. However, if the aim is to generate as many change actions as possible, you can list all constructs in which the Website to be improved is either rated as positive ("+") or negative ("−"). This larger list of constructs can also be grouped into topics according to the procedure previously described, which in turn will be the starting point for generating change actions as describe in the present section.

CONCLUSION

In the present chapter, I have proposed the Character Grid based on the Repertory Grid Technique as a method for exploring and evaluating Website designs. It certainly has its limitations. One of the limitations to be aware of is the lack of structure to the gathered construct pool. The exact relationships among constructs are hard to determine based on the grid data alone. It may be a causal relationship, such as "the site has an 'sober' appearance, which makes it suitable 'for business people,' " simple co-occurrence, such as "the site has a 'neutral color' scheme and its 'target user group is vague,' " or a hierarchical relationship, such as "the site is generally 'news oriented,' which comprises of having 'current news on home page' and a 'news ticker.' " A good deal of reasoning and interpretation is required to understand relationships between constructs. Nevertheless, it seems worth a try if user perceptions, needs, beliefs, and attitudes are to be taken into account by a Website design.

Beyond questions about the mere practicality of the Character Grid, the underlying personal construct approach addresses a more fundamental issue, namely the informational value of the ideographic (i.e., personal) perspectives of users on the Websites we design. Human–computer interaction research tends to generalize across users. This view—rooted in the quantitative research tradition—neglects the informational value of contradictions and inconsistency in ideographic perspectives. Moreover, the field of usability engineering tends to undervalue what people experience and report. Subjectivity seems to be something to avoid. However, subjectivity matters when users have a personal choice, and they have a choice even in the more restricted work environment. Igbaria, Schiffmann, and Wieckowski (1994), for example, showed how perceptions of usefulness and enjoyment of the computer in the workplace is being reflected in the time spent and the number of different business tasks done with the computer.

To conclude, I strongly believe that much can be learned from the naïve theories people have about the artifacts they live with and the differences among and contradictions within these theories.

FURTHER READING

The Character Grid presented in this chapter is a variant of the original Repertory Grid Technique. It describes a procedure and analysis techniques I found helpful for guiding the design of Websites. However, Fransella and Bannister (1977) stress that "the grid is truly a technique and one which is only limited by the user's lack of imagination" (p. 59). Thus, I want to encourage you to further adapt the Character Grid to your special needs.

A good start for learning more about the Repertory Grid Technique is Fransella's and Bannister's (1977) book *A Manual for Repertory Grid Technique*. Although not of recent date, it gives a good overview of the technique and its variants. A very excellent, more recent, source of information about the Repertory Grid Technique is Gaines' and Shaw's (2001) Website (http://gigi.cpsc.ucalgary.ca/). They even offer "WebGrid III," a free Repertory Grid tool that allows a computer-supported elicitation and element rating procedure and a variety of statistical analyses.

ACKNOWLEDGMENTS

I would like to thank Julie Ratner, David Forrester, and Kim Daumiller for reviewing the present chapter, Uta Sailer for her very helpful comments on earlier drafts, Tibor Trautmann for collecting the best part of the data used as examples, and the people at User Interface Design GmbH for their support.

REFERENCES

Baber, C. (1996). Repertory grid theory and its application to product evaluation. In P. Jordan, B. Thomas, B. A. Weerdmeester, & I. L. McClelland (Eds.), *Usability evaluation in industry* (pp. 157–165). London: Taylor & Francis.

Bellotti, V. (1988). Implications of current design practice for the use of HCI techniques. In D. M. Jones & R. Winder (Eds.), *Proceedings of the HCI '88 Conference on People and Computers IV* (pp. 13–34). Cambridge, England: Cambridge University Press.

Bevan, N., & Macleod, M. (1994). Usability measurement in context. *Behaviour & Information Technology, 13*, 132–145.

Dillon, A., & McKnight, C. (1990). Towards a classification of text types: A repertory grid approach. *International Journal of Man–Machine Studies, 33*, 623–636.

Fransella, F., & Bannister, D. (1977). *A manual for repertory grid technique.* London: Academic.

Gaines, B. R., & Shaw, M. L. G. (1997). Knowledge acquisition, modelling and inference through the World Wide Web. *International Journal of Human–Computer Studies, 46*, 729–759.

Gaines, B. R., & Shaw, M. L. G. (2001). Webgrid III [Computer software]. Available: http://gigi. cpsc.ucalgary.ca/ (Last visited November 6, 2001).

Gould, J. D., & Lewis, C. H. (1985). Designing for usability: Key principles and what designers think. *Communications of the ACM, 28*, 300–311.

Hassenzahl, M., Burmester, M., & Beu, A. (2001). Engineering joy. *IEEE Software, 1/2*, 70–76.

Hassenzahl, M., & Trautmann, T. (2001). Analysis of web sites with the Repertory Grid Technique. In M. Beaudauin-Later & R. J. K. Jacob (Eds.), *Proceedings of the CHI 2001 Conference on Human Factors in Computing. Extended abstracts* (pp. 167–168). New York: Association for Computing and Machinery, Addison-Wesley.

Hassenzahl, M., & Wessler, R. (2000). Capturing design space from a user perspective: The Repertory Grid Technique revisited. *International Journal of Human–Computer Interaction, 12*, 441–459.

Igbaria, M., Schiffman, S. J., & Wieckowski, T. J. (1994). The respective roles of perceived usefulness and perceived fun in the acceptance of microcomputer technology. *Behaviour & Information Technology, 13*, 349–361.

International Standards Organization. (1999). *ISO 13407: Human-centred design processes for interactive systems*. Geneva: Author.

Janlert, L.-E., & Stolterman, E. (1997). The character of things. *Design Studies, 18*, 297–314.

Kelly, G. A. (1955). The psychology of personal constructs (Vols. 1 & 2). New York: Norton.

Logan, R. J., Augaitis, S., & Renk, T. (1994). Design of simplified television remote controls: A case for behavioral and emotional usability. In *Proceedings of the 38th Human Factors and Ergonomics Society Annual Meeting* (pp. 365–369). Santa Monica, CA: Human Factors and Ergonomics Society.

Monö, R. W. (1997). *Design for product understanding: The aesthetics of design from a semiotic approach*. Stockholm: Liber.

Moody, W., Morgan, R., Dillon, P., Baber, C., & Wing, A. (2001). Factors underlying fabric perception. In C. Baber et al. (Eds.), *Proceedings of the Eurohaptics 2001 conference* (pp. 192–201). Birmingham, UK: University of Birmingham.

Moran, T. P., & Carroll, J. M. (1994). *Design rationale: Concepts, techniques, and use*. Hillsdale, NJ: Lawrence Erlbaum Associates.

Osgood, J. H., Suci, G. J., & Tannenbaum, P. M. (1957). *The measurement of meaning*. Urbana: University of Illinois Press.

Sinclair, M. A. (1990). Subjective assessment. In J. R. Wilson & E. N. Corlett (Eds.), *Evaluation of human work* (pp. 69–100). London: Taylor & Francis.

Slater, P. (1976). *The measurement of intrapersonal space by grid technique: Vol. 1. Explorations of intrapersonal space*. London: Wiley.

11

Eye–Hand Coordination During Web Browsing

Mon-Chu Chen

Carnegie Mellon University,
Human–Computer Interaction Institute

John R. Anderson
Myeong-Ho Sohn

Carnegie Mellon University,
Psychology Department

INTRODUCTION

Web designers seek to provide optimal user experiences. Traditionally, the technique of observing users has been employed to help designers solve usability problems. By extending our understanding to the user's cognitive processes during Web browsing, we could additionally enhance users' experiences (e.g., by providing adaptive services). To date, the most common method for understanding users' behaviors has been to record the users' activities during Web browsing to a log file. A log file consists essentially of a set of time stamps and the URLs a user selected. Some research has tried to push the logged data further. For instance, in addition to URL selection, Farrell (1999) tried to capture the interaction between users and Web server in a finer granularity by looking at all users' inputs and system responses, such as URL selections, FORM inputs, plug-in and applet activities, and others. Most recently, however, mouse movement itself has been examined as a source of information on which to model users' behaviors (Chen, Anderson, & Sohn, 2001; Mueller & Lockerd, 2001).

Another technique to monitor users' behaviors is eye tracking (Marshall, 2001). This technique has been used as a powerful tool in various psychology experiments. For example, the Poynter Institute (2000) conducted a study of users reading news

Websites while their eye movements were monitored. One interesting finding of this research is that graphics do not grab users' visual attention at the beginning of a page. This finding could not have been determined without use of an eye tracker. Other companies, such as EyeTools, Inc. (Newark, California) and EyeTracking, Inc. (San Diego, California), have even started to provide commercial services to improve Website design using eye-tracking techniques.

To date, most studies have examined either mouse events and movements or eye-tracking data. Because these measures have not been examined simultaneously, we have limited information about the interaction between mouse and eye movement. The present experiment was designed to study the interactions between users and the computer system by combining these two measures. Eye–hand coordination is essential for human–computer interactions. The eye measure describes how a participant perceives information in a Web browser. The mouse measure evaluates how a participant responds to the computer system using the mouse cursor. In the following sections, patterns of eye and mouse movements during the context of Web browsing are presented. The implications and applications of the findings are also discussed.

EXPERIMENT DESIGN

System Setup

An EyeLink eye-tracking system (SensoMotoric Instruments, Inc., Boston, Massachusetts) was installed on an IBM-compatible PC (referred to as the tracking PC). The EyeLink system is a video-based, binocular eye-movement tracking system. Two cameras were attached to the headband to capture eye images. A third was used to track the four infrared emitters attached to the display to track the head position. Images from three cameras were processed by the tracking system to calculate the gaze positions, eye saccades, and eye fixations. A separate PC (the experiment PC) was used by participants to view Websites. These two machines were connected via an Ethernet connection in order to transmit mouse data and Web events as well as commands for controlling the eye tracker. In order to reduce the effect of network traffic on the connection between the tracking PC and the experiment PC, these two machines were isolated from the other machines on the campus-wide network. Website files were downloaded and stored on the local-experiment PC hard drive. During the experiment, the cached Web pages were accessed directly from the hard drive in order to reduce the possible variations in page loading time across participants. Storage space prohibited the caching of entire Websites. Therefore, a Website was mirrored up to three levels deep from the home page of that Website. All external links and links to uncached pages were disabled.

A customized Web browser was developed in Visual Basic for this experiment. In addition to the functions of a regular browser, the program could communicate with the tracking PC and remotely control the eye tracker. Users' browsing activities were also recorded in the form of events and their time stamps. Moreover, the browser recorded both the mouse button events and movements. The browser contained special low-level codes, which were written to have equivalent sampling rates for the mouse movements and eye movements (250 Hz). However, due to the limitation of the Windows 98 operating system, the sampling rate of the mouse movements was not always a constant (e.g., when accessing the hard drive or redrawing the screen). Therefore, an interpolation process on the mouse data was done.

Experimental Procedures

Participants were asked to wear the eye tracker during the experiment. The eye-tracking system was calibrated at the beginning of the experiment. Participants were instructed to explore a particular Website for 5 minutes. A drift-correction process was conducted to maintain tracking accuracy before the next Website was viewed. In some cases, the system required recalibration between viewing different Websites. Participants explored four Websites, each with a different visual style in terms of layout, color scheme, and information architecture. The original URLs for the sites were

- www.cmu.edu (a multifunctional Website serving a large university community, thus containing Web pages with various styles).
- www.cmoa.org (a museum site with a consistent style of page layout and with many graphics).
- www.apple.com (a company Website with similar page style and different functions, such as shopping/portal/info).
- www.sapient.com (a company Website consisting mainly of information about the company with a consistent layout style).

These sites were selected to give us some ideas of how consistent our results were over different styles of Websites.

During the experiment, the same order for visiting Websites was maintained across all participants. Participants were told that their eye movements would be monitored. It was not mentioned that their mouse activities would be monitored as well so that their mouse actions would not be affected by such knowledge. Users could stay on a particular page for as long as they liked. Users could click any link in any order as long as the destination page was one of the cached pages. When the cursor was over the link for a destination page that was not cached, the cursor would become a stop sign to indicate that link was disabled. After 5 minutes, users were allowed to take a rest. The total experimental session usually took less than 40 minutes to complete.

Data from 5 participants were collected. All 5 participants were recruited from a university setting. Their ages ranged from 25 to 30, and 2 of the 5 participants were female. During the session, the log files and mouse data were transmitted in real time over the Ethernet connection to the tracking PC and were merged with the eye-movement data into a single data file. The time-stamped data set contained information about gaze position, pupil size, cursor position, Web page information, and mouse button status. A program was developed to parse eye saccades, eye fixations, as well as mouse saccades and mouse fixations. The default setting of the eye tracker was used for categorizing eye movements into fixations and saccades. The criterion for mouse fixation was as follows: The speed of the cursor was slower than 75 pixel per second, and the cursor stayed within a circle with a radius of 10 pixels for at least 1.2 seconds. Mouse movements other than that were treated as mouse saccades.

RESULTS

Because of the free-form navigation, each participant visited a different set of pages. A page was operationally a page of certain Website; a visit was defined as an instance of a loaded page during the navigation.

After a user clicked a link, there was a brief latency while the browser loaded the page and rendered the contents on the screen. Our definition of a visit is an instance of a loaded page starting from the time when the page and all of its components (e.g., frames) were fully loaded until the time when the mouse button was pressed to leave the page (e.g., another link selected). To simplify data analysis, we only analyzed data from the initial view of the page in the browser (i.e., that which was visible without scrolling downward).

Table 11.1 shows the number of visits by each participant of each Website. Table 11.2 shows the number of unique pages visited by each participant of each Website. In total, 595 visits were made to 235 different Web pages. The reason why there are more visits than pages is because participants may have visited a particular page more than once. The overall number of pages does not equal

TABLE 11.1
Number of Visits

Site	Participant 1	Participant 2	Participant 3	Participant 4	Participant 5	Number of Visits
www.apple.com	15	26	36	13	30	120
www.cmoa.org	31	30	59	11	25	156
www.cmu.edu	33	49	46	15	30	173
www.sapient.com	31	18	62	12	23	146
Total	110	123	203	51	108	595

TABLE 11.2
Number of Pages

Site	Participant 1	Participant 2	Participant 3	Participant 4	Participant 5	Number of Pages
www.apple.com	11	12	31	5	14	44
www.cmoa.org	15	22	33	6	15	56
www.cmu.edu	17	21	33	9	17	72
www.sapient.com	23	17	54	7	22	63
Total	66	72	151	27	68	235

the sum of the individual number of pages of each participant because different participants may have visited the same unique page. Also, the number of visits and pages varied across participants.

Examples of Mouse and Eye Traces

As discussed in the section above, both eye movements and mouse movements have been used independently to describe users' intentions. Here, by comparing both the eye and the mouse movements, we are able to infer a richer understanding of the user's cognitive processes during Web navigation. Figure 11.1 shows a series of snapshots of a typical eye movements and mouse movements overlaid on a particular Web page. The dark line indicates the path of the mouse cursor. The gray line indicates the scan path of the user's eyes. The dots indicate the fixation point of the eyes and mouse cursor, respectively.

Each snapshot in Fig. 11.1 represents the progression of the navigation in approximately a 1-second interval. In (a), both gaze and cursor started from the top-left corner of the screen. The participant quickly shifted her gaze to the center of the page and then moved to the menu panel on the left side. Meanwhile, the cursor was moved slowly toward the center of the page. In (b), the cursor finally was fixed at the picture with a portrait at the center of the page. At the same time, the participant visually scanned from the menu panel on the left, across the mouse path, fixated on the text box on the right side, and then scanned back to where the cursor was fixed. In (c), the cursor was moved to the place where the user had been looking, which was one of the hyperlinks. However, it can be inferred that the user was not interested in that link anymore because the user's gaze shifted to the menu bar on the left-bottom corner and let the cursor rest there. In (d), the user continued scanning the screen from the left side to the text in the bottom-right corner and then down to the banner on the bottom. While paying attention to the banners, the user moved the cursor slowly toward to the bottom-right corner. However, before the mouse cursor reached the banner, we can see in (e) that the user checked the banners and then returned to the menu panel

(a)

(b)

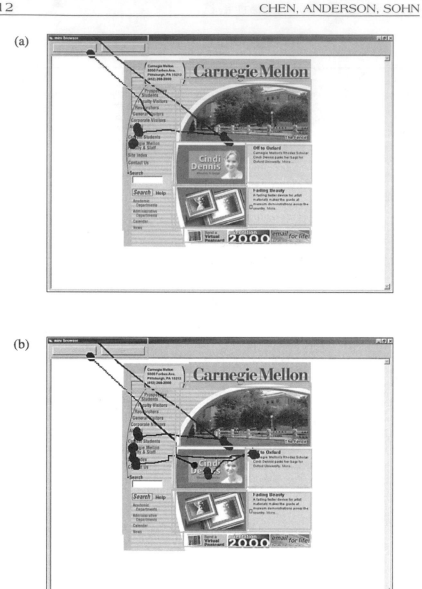

FIG. 11.1. Snapshots of eye/mouse movements on a visit.

(c)

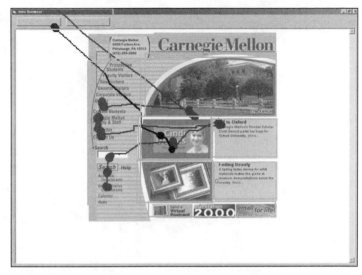

(d)

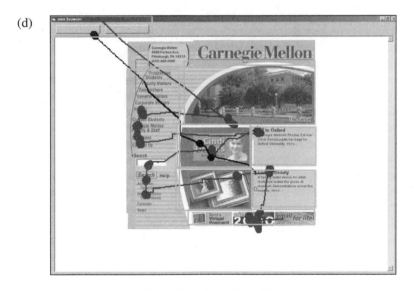

FIG. 11.1   (Continued)

(e)

(f)

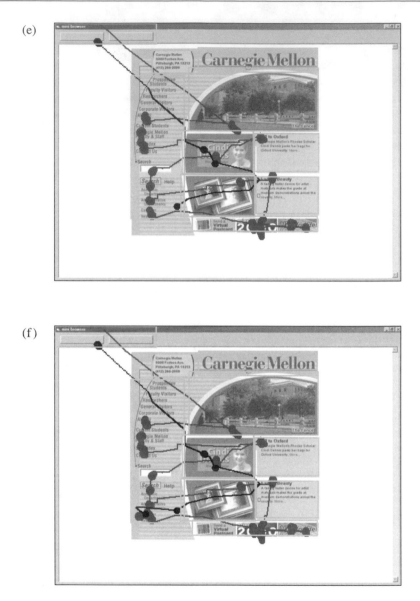

FIG. 11.1   (Continued)

again. In the meantime, the user suddenly changed the direction of the mouse saccade and turned 120 degrees clockwise, toward the menu panel. This was a clear sign that the user was not interested on the banners on the button-right corner. Finally, in (f), the user made some adjustments and clicked one of the menu items.

This example illustrates the power of combining mouse movement and eye-movement data. The entire navigation process could not be easily interpreted by relying solely on one of these measurements. In the following, we present several specific examples to illustrate some combinations of eye and mouse movements.

Busy Eyes and Lazy Mouse

In Fig. 11.2(a), we see a clear pattern of eye movements associated with reading. This pattern continued for approximately 9 seconds. During this period, the mouse cursor was stationary. Without the eye-tracking data, the participation's cognitive status would have been unknown.

Change of Interest in What Is Seen

One general idea in adopting eye tracking for Web design evaluation is to use users' visual attention to infer their interests on the Web. However, Fig. 11.2(b) demonstrates a case in which the user moved the cursor toward the highlighted square on the middle-right of the page, where the user was looking. Just before the cursor reached the square, the user returned the cursor to its original location— the menu items on the top-left corner. Without the mouse data, we may infer that the participant was interested in that region. However, the mouse data further suggests that the user lost interest in that region and decided not to do any further action (e.g., click that link). Therefore, with eye data only, we cannot have information about the decisions that users made, although we can tell where users were looking.

Mouse Follows Gaze

Most of the eye- and mouse-movement patterns were similar to the one shown in Fig. 11.2(c). The mouse cursor generally followed the path of the eye movements. Both movements shared a similar path and visited a similar set of regions. In this case, eye gaze and mouse cursor paths coincided frequently.

What More Can a Mouse Cursor Tell Us?

In the case of (d), the data show a cluster of eye fixations and mouse fixations on the block above the center of that page. Interestingly, the user moved the cursor diagonally downward away from the block and then moved it back to the block.

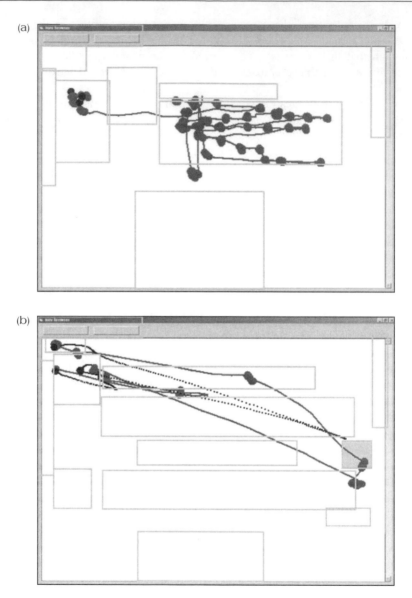

FIG. 11.2   Examples of eye/mouse movements.

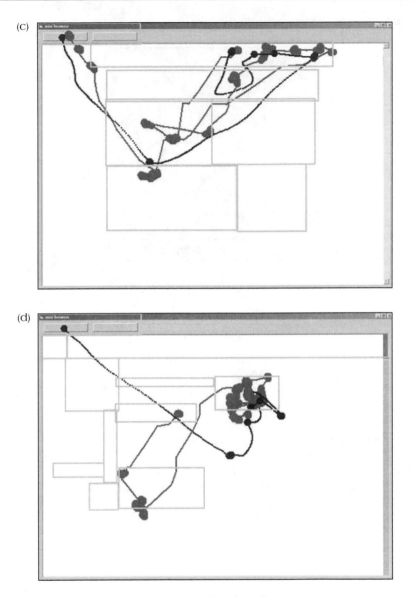

FIG. 11.2   (Continued)

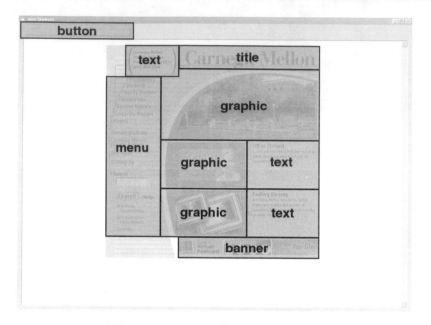

FIG. 11.3 Regions of a particular page.

This short diagonal mouse movement originating from some informative place indicated that the user may have been looking at that place and needed to move the mouse cursor in order not to block where he wanted to look. This example indicated the possibility of predicting gaze position from mouse movements without using an eye tracker.

Another example that showed a good opportunity to predict gaze is in Fig. 11.2(a). At the end of this visit, several consecutive short mouse saccades falling on several small elements (e.g., menu items), implied that the user was selecting from several options. This process required the user to look at that location to direct the cursor toward the desired selection. Therefore, it was very likely the eye gaze would be in this specific area.

DATA ANALYSIS

For the purpose of data analysis, several regions were defined for each page. A region was defined as a unique Web element or a set of like elements (e.g., menus, text, graphics, and banners). Areas that did not contain any of these elements were categorized as "nowhere regions." Regions for each of the 235 unique pages were defined by the first author prior to data analysis. Given that each page was unique, the number and size of regions varied across pages. Figure 11.3 shows an example of a set of regions for a particular page.

TABLE 11.3

Ratio of Regions Visited by Eye/Mouse

	Not Visited by Eye (E0)	Visited by Eye (E1)
Not visited by mouse (M0)	0.44	0.18
Visited by mouse (M1)	0.06	0.32

TABLE 11.4

Correlation of Eye/Mouse Dwelling Time

	r
www.apple.com	0.59
www.cmoa.org	0.82
www.cmu.edu	0.72
www.sapient.com	0.41

In each visit, all regions were assigned one of four visit status categories: visited by both mouse cursor and gaze (E1M1), visited only by mouse (E0M1), visited only by gaze (E1M0), and visited by neither mouse nor gaze (E0M0). For each visit, the proportion of each category of visit status was calculated. Table 11.3 shows the mean proportion over all 595 visits. It is evident from these data that the majority of regions (44%) were never visited by either the mouse or gaze (E0M0) within a particular visit. The second-most frequent visit status category was E1M1 (visited by eye and mouse). This implies that when the mouse cursor visits a certain region, the probability of that region being visited by the gaze is very high, $P(E1/M1) = 0.84$. Moreover, when the mouse cursor does not visit a certain region, the probably of that region not being visited by gaze is also high, $P(E0/M0) = 0.71$. Regions visited by the mouse but not by the gaze are rare (6%).

The above analysis does not address how long an eye gaze or a mouse cursor stayed in a particular region. Since the dwell time of an eye gaze at a particular area is used to infer users' interests in that area, it is important to examine the correlation between the dwell times of eye gaze and mouse cursor. In each region of each visit, we calculated the total time for both the gaze and cursor dwelling in that particular region; the overall correlation of these values was $r = .58$. Table 11.4 shows the range of correlations for individual Websites; the range of correlation values was .41 to .82. The observed range of values may be related to layout within a particular Website. For example, the www.cmoa.org site has a consistent layout of pages across the site and is more graphic intensive than others. This type of design may require more eye–hand coordination for participants to interact with the Website, leading to a greater correlation between gaze and cursor dwelling times.

So far, the relationship between eye gaze and mouse–cursor movement has been addressed in the global sense. Now, we move forward to discuss individual mouse activity and corresponding eye behaviors by calculating a gaze-meet-cursor score,

TABLE 11.5
Gaze-Meet-Cursor Score and Gaze–Cursor Distance

Mouse Events	Gaze-Meet-Cursor Score		Gaze–Cursor Distance	
	M	SD	M	SD
Mouse fixation (MF)	0.64	0.12	73.41	17.53
Mouse saccade beginning (MSB)	0.50	0.09	148.11	36.01
Mouse saccade ending (MSE)	0.61	0.12	104.08	33.71

described later. First, three mouse events were defined: whole mouse fixation (MF), the beginning of a mouse saccade (MSB), and the end of a mouse saccade (MSE). Again, the mouse fixation was defined as follows: the speed of the cursor was slower than 75 pixels per second, and the cursor stayed within a circle with a radius of 10 pixels for at least 1.2 sec. MSB was defined as the period from the beginning of the mouse saccade till the moment the cursor left the region (with a maximal duration of 0.1 second). MSE had a similar definition to MSB except with MSE ending at the completion of that mouse saccade. During a mouse event, if eye gaze visited the same region as that in which the mouse cursor resided, the gaze-meet-cursor score was coded as 1 regardless of how many times the eye gaze visited that region. Note that only the mouse events, which occurred in a meaningful region (i.e., a non–nowhere region), were counted in this analysis. Table 11.5 shows the means and standard deviations for the gaze-meet-cursor scores for the three different types of mouse events. These numbers convey the chance that a region will be visited by the eye gaze during a particular mouse event. The data show that when the mouse cursor is fixed in a meaningful region there is a 64% chance that the eye gaze will visit the same region during the mouse fixation. The data also suggest that the chance for the eye gaze to meet the mouse cursor at the end of a mouse saccade is higher than at the beginning.

The distance between the eye gaze and the mouse cursor was also measured. During an MF, the gaze–cursor distance was measured as the distance between the average cursor position and the average position of eye fixations within the same region. While in MSB and MSE, the gaze–cursor distance was measured as the minimum distance between gaze and cursor within the window. Table 11.5 shows the data for each type of mouse events. The data show that when the cursor was fixed in a meaningful region, the eye gaze was very close to the cursor position. The overall mean distance was 73.41 pixels, which is less than 3 degrees of visual angle. A significant difference was found between the gaze–cursor distance during MSE and that observed during MSB ($F = 5.31, df = 1, 12, p < .05$). That implies that the gaze was closer to the cursor during MSE than during MSB. In other words, a mouse saccade was moving toward the gaze position.

The last analysis presented here is different from the previous ones. This analysis was called mouse fixation transition. It examined the relationship between the gaze

TABLE 11.6
Mouse Fixation Transition

		Gaze-Meet-Cursor Score (of Next MF)	
		NO	N1
Gaze-meet-cursor score	C0	0.19	0.17
(of current MF)	C1	0.13	0.51

and the cursor over a period of time. A mouse fixation transition consisted of two sequential mouse fixations. By adopting the gaze-meet-cursor score previously described to the concept of mouse fixation transition, we defined four types of transitions:

1. The gaze did not meet the cursor during this MF and would not meet during next MF (C0N0).
2. The gaze met the cursor during this MF and next MF (C1N1).
3. The gaze met the cursor during the first MF but not for next MF (C1N0).
4. The gaze met the cursor during the next MF but not for the current MF (C0N1).

The percentage of each transition is shown in Table 11.6. The C1N1 is the most frequent type of transition. When the eye gaze meets the mouse cursor during a mouse fixation there is a very high chance that the eye gaze will meet the cursor again in the next mouse fixation $P(N1/C1) = 0.8$. This suggests that very often the mouse cursor shares the same path as the eye gaze.

DISCUSSION

Implications

When using personal computer systems, people use their eyes to search the visual display and then employ their hand to move the mouse cursor in order to issue commands. This eye–hand coordination is essential for people to interact with a computer system. It is reasonable to assume that the relationship between the eye gaze and the mouse cursor will vary under different contexts. For example, one might expect stronger gaze/cursor relationships in drawing software such as Photoshop than in a text editor such as Word. Different computer systems may require different levels of eye–hand coordination. In the realm of the World Wide Web, the density of visual information and clickable links make Web browsing a relatively high interactive task. Therefore, the relationship between the eye and the mouse during Web browsing becomes an interesting issue to address.

By examining eye data, we can understand what participants perceived during Web browsing. By looking at mouse data, we can understand how participants react to the system. By combining eye and mouse traces, we can better understand users' behavior and more precisely interpret users' intentions.

The strong correlation found in this study between eye movements and mouse movements suggest that mouse data can represent the eye data on Web browsing quite well. For those Web developers who do not have any access to an expensive eye tracker, a mouse tracker can serve as a reasonable alternative. The cursor can tell us not only about the cursor position on the screen but also about where the users are looking and reveal their inner cognitive status under certain conditions (e.g., choosing between several options, as seen in Fig. 11.2[b]).

Applications

The use of eye trackers in Web design evaluation is not a new idea. However, an eye tracker is expensive, such that most people will not be able to access this type of equipment. By employing a mouse tracker, we could build a user model without the use of an eye tracker. Because the mouse is the most widely used input device for Web browsing, it is technically practical to construct a simple browser plug-in or a special script embedded into a Web page to remotely collect users' behavioral data over the Internet.

Employing a mouse tracker on a Website will benefit individuals doing research on Web development and users' studies. The results could also help designers provide an improved experience for visitors to their Websites. This study suggests that a mouse cursor can convey to us more than the just X and Y coordinates. There are potentials of providing new interactive techniques based on the inferred users' cognitive status.

The findings of this study may be applicable to other human–computer interactions beyond the Web browser. Whenever an eye tracker is unavailable, a mouse tracker may be used as a replacement to certain degree. Future studies are needed to confirm this relationship on systems other than Web browsers.

Limitations

The data suggest that the mouse is less active than the eye during Web browsing. Whenever the mouse is stationary, we cannot infer the activities of the user's eyes. The context may provide us with some clues to make a good guess. However, consider the example in Fig. 11.2(a), where the mouse cursor is fixed while the eye gaze pattern shows that the participant was reading a paragraph. Under this condition, we may say that the gaze is somewhere other than the place that the cursor dwells, but there is no way to infer gaze position under this situation.

For remote data collection, bandwidth may be an issue because of the demand of continuously transmitting mouse activities over the Internet. To reduce network traffic, analyzing the mouse data locally in the participant's machine might be possible. However, this may require more complicated software development to achieve this goal.

One other important issue is that of user privacy. Is it legal and ethical to collect users' behavioral data in this way without noticing them? If we ask their permission at the beginning, there will be a great chance that people will not allow this or may alter their mouse behaviors unconsciously.

Due to several technical issues of implementing such a system and processing the huge amount of the data set, we were only able to acquire the data of 5 participants. We expect to test more participants in the future to determine the external validity of the results we found.

CONCLUSIONS

This chapter described an experiment in which participants' eye movements and mouse movements were recorded as Websites were browsed. The combined eye and mouse traces revealed interesting patterns of eye–hand coordination during Web browsing. By combining eye and mouse protocols, we may be better able to understand users' intention and decision-making patterns. The data analysis showed that there was a strong correlation between the two measures. When an eye tracker is unavailable, it is possible to use a mouse tracker to help collect data on users that traditional Web logging methods cannot acquire.

ACKNOWLEDGMENTS

Mon-Chu Chen would like to express special thanks to Dr. Valerie Monaco for her assistance in editing the text of this chapter.

REFERENCES

Chen, M.-C., Anderson, J. R., & Sohn M.-H. (2001). What can a mouse cursor tell us more? Correlation of eye/mouse movements on web browsing. In J. Jacko & A. Sears (Eds.), *Human factors in computing systems: Extended abstracts of CHI '01* (pp. 281–282). Seattle, WA: Association for Computing and Machinery.

EyeTools, Inc. (2001). Uses innovative patented eye-tracking technology developed at Stanford University to produce dramatic improvements in web site design [online]. Available: http://www.eyetools.com (Accessed November 16, 2001).

EyeTracking, Inc. (2001). EyeTracking offers Website and software usability studies using eyetracking technology [online]. Available: http://www.eyetracking.com (Accessed November 16, 2001).

Farrell, R. (1999). Capturing interaction histories on the Web. In *Proceedings of the Second Workshop on Adaptive Systems and User Modeling on the WWW, the Eighth World Wide Web Conference* [online]. Available: http://www.research.ibm.com/Applied LearningSciWeb/Farrell/Adaptive.html. Toronto, Canada:

Marshall, S. (2001). Eye tracking: A rich source of information for user modeling. In M. Bauer, P. J. Gmytrasiewicz & J. Vassileva (Eds.), *User Modeling '01 Conference proceedings* (p. 315). Sonthofen, Germany: Springer-Verlag.

Mueller, F., & Lockerd, A. (2001). Cheese: Tracking mouse movement activity on Websites, A tool for user modeling. In J. Jacko & A. Sears (Eds.), *Human factors in computing systems: Extended abstracts of CHI '01* (pp. 279–280). Seattle, WA: Association for Computing and Machinery.

Poynter Institute. (2000). Stanford Poynter eyetrack project: Study of reading of on-line news site [online]. Available: http://www.poynter.org/eyetrack2000/ (Accessed November 16, 2001).

12

Designing Friction-free Experience for E-commerce Sites

Pawan R. Vora
Seurat Company

INTRODUCTION

Despite the recent downturn in the economy and world events, U.S. retail e-commerce sales for the fourth quarter of 2001 totaled $10.043 billion according the Department of Commerce, up 13.1% from the fourth quarter of 2000. Furthermore, according to Jupiter Media Metrix (2001), on average, 51.3 million unique visitors went to shopping sites each week during the 2001 holiday-shopping season, up 50 percent compared with the 2000 holiday-shopping season and up 95 percent versus 1999. Jupiter Media Metrix also expects that this trend will continue and consumers will increasingly allocate a greater percentage of their holiday budget to online shopping.

With more and more sites becoming e-commerce enabled, consumers will have several choices for making purchases in any of the retail categories. What will make consumers favor one e-commerce site over the others? According to the research by GartnerG2, 81% of online consumers value convenience when making a purchase online, compared with 33% who value price savings. The reason convenience is favored over price is because typical Internet consumers live a wired lifestyle and are time deprived (Bellman, Lohse, & Johnson, 1999) and consequently have a limited amount of discretionary time. Therefore, e-commerce sites should focus create a customer experience where customers can achieve their goal of purchasing products or services efficiently and effectively.

In this chapter, the notion of friction-free experiences is proposed as a way to create such a compelling and convenient customer experience.

FRICTION-FREE EXPERIENCES: A DEFINITION

The following story about an annual meeting of a major manufacturer of power tools explains the notion of friction-free customer experience very effectively (Hammer, 2001):

> The chairman stands up to address the assembled shareholders. "I have some bad news for you," he says. "Nobody wants our drills." The audience is shocked. At last report, the company had a 90 percent share of the drill market. The chairman proceeds, "That's right, nobody wants our drills. What they want is holes."

As in the above story, where the company's success is dependent upon how well it helps its customers achieve their goal of making holes, the success of an e-commerce site depends on how well it helps its customers achieve their **goal of purchasing a product or a service**. Friction refers to the activities and tasks that customers are required to do that are extraneous to their goal of purchasing.

For a bricks-and-mortar store, the activities that can be considered friction are: traveling to the store, parking the car, traveling from the parking lot to the store, finding the product in the store, time spent in the checkout line, and paying for the purchase (see Bhatnagar, Misra, & Raghav Rao, 2000). All these activities require investment of time; however, they have nothing to do with the goal of purchasing the product. Purchasing the same product, if available, on the Internet helps eliminate the previously defined friction and allows placing an order from the confines of one's home. In fact, it is this convenience that has made e-commerce enticing for many consumers; similar conveniences have made catalog shopping, telephone shopping, and TV shopping attractive as well. However, purchasing online is not entirely friction-free. It introduces friction of its own, such as time spent browsing Websites (which includes the page download times), finding the desired product on sites (especially when the navigation is cumbersome), checking out the items in the online shopping cart, waiting for the product to arrive at your doorstep, and, when necessary, returning the product or dealing with the customer service representatives for delayed or unshipped orders.

Do we really need to eliminate or minimize these frictions? Yes. Because, as discussed earlier, the prototypical Internet customers are time deprived and that e-commerce sites should focus on getting the customer in and out of the site as quickly and efficiently as possible. That is, they should attempt to create a friction-free customer experience.

Types of Friction in Online Shopping Experience

The types of friction that exist for online customers could be mapped to the most common tasks customers engage in when shopping (Kalakota & Whinston, 1996).

- **Prepurchase friction:** e.g., search for products, compare product characteristics, negotiate transaction details, etc.
- **Purchase friction:** e.g., place items in the shopping cart, checkout, get payment authorization, place the order, etc.
- **Postpurchase friction:** e.g., track shipping progress, return a product, contact customer service representative, etc.

The goal of the customer experience architect of an e-commerce site then is to design Web-based interfaces that either eliminate or compress the time required to perform these tasks. In the following sections, I will offer guidelines for designing friction-free customer experiences for e-commerce sites and explain how they can offer substantial efficiency benefits to their customers.

GUIDELINES FOR DESIGNING FRICTION-FREE EXPERIENCES

Think: Customer

The main difference between designing interfaces for users and customers is that of the designer's mind-set. Designing for the latter requires the perspective not of users of computer systems, but of customers engaged in commercial transactions (Lee, Kim, & Moon, 2000). Then, like real world commerce, the objective of the design is not just to design efficient and effective experience for first-time customers but also for repeat customers. Thus, the designers have to be concerned not only with converting browsers into buyers during their earlier visits but also with keeping the customers returning for future purchases and making them loyal customers. This is not to suggest that they are not interrelated—for example, there is evidence that customer loyalty is largely determined by how quickly users are able to find what they are looking for on an e-commerce site—but to acknowledge that it is those repeat visits where substantial time-saving benefits can be offered to the customers and that the system architecture should be designed accordingly.

Removing Prepurchase Friction

Get Users to the Desired Product Quickly

The very first task when arriving at an e-commerce site—after getting oriented with what the site has to offer—is to find the desired products and/or services. The common design choice to help users in this endeavor is to design elaborate navigation schemes. However, as Cooper (2001) pointed out, many Web designers have incorrectly deduced that users *want* navigation schemes: "Once a beginner's enthusiasm wears off after a few uses of the Web, she would just like to get her work done in the simplest and most straightforward way possible. Instead of building a complex structure of pages, a better design technique is to concentrate all of the interaction in a single screen, relieving the user of the need to explore, or the need to navigate at all." Although Cooper seems to suggest an extreme scenario of not having a navigation scheme at all, which is often not feasible, it is important to understand that navigating or searching is essentially a friction or an obstacle to the user goal of buying a product or a service. Helping users find the product they want quickly can reduce prepurchase friction. A good design should help users decide on a product and also assure them it's the product they want. Here are some approaches to reduce prepurchase friction:

• **Merchandising or promotions.** Lohse and Spiller (1998b, 1998a) showed that promotion on the home pages could increase both visitor traffic and sales. Why? Because promotions directed customers quickly to an individual product, eliminating the need to navigate and thus reducing the friction.

• **Improved category links.** We cannot read customer's minds, though. There will be instances when customers know what they want and will use navigation and search mechanisms provided on the sites. Research by User Interface Engineering (2001b, 2001a) showed that the major driver of impulse purchases was the use of the category links on the site. When category links were designed keeping customer goals in mind, the customers were motivated to find products through category links rather than the sites' search engines, and when using the category links, they were far more likely to make impulse purchases than when they used search engines. Simply having category links is not enough, however. Category links should be designed to meet customer's need, and multiple ways to get to a product may need to be offered. For example, for an electronics store, category links should not only be based on product types (home audio, digital cameras, etc.), but also be based on brand (e.g., Sony, Panasonic, RCA, etc.) as well as price (e.g., less than $100, $100–$200, etc.).

• **Improved product lists.** When encountering product lists when navigating or searching, it is important that information such as price, a thumbnail image, and longer descriptive product names are shown in the lists and not force the customers to go to a Product Details page to get such information. As shown by Lohse and Spiller (1998b), product lists had the largest impact on sales and it accounted for 61% of the variance in monthly sales.

All of the above can be categorized as means to improving navigation, which is useful not only for helping customers get to the product quickly, but also is a necessary precondition to successful communication of a site's trustworthiness, as demonstrated in the studies by Cheskin Research (2000; Cheskin Research & Studio Archetype/Sapient, 1999).

Help Users Make Purchasing Decisions

When customers are not experts in the product domain, the previously described approaches may not work. For example, a customer may want to buy a digital camera or a DVD player but may not know exactly what he or she should buy. To help such customers make correct purchasing decision, e-commerce sites should help them find the right products that match their needs. Here are few ways to do so:

• **Offer Shopping Guides, Recommenders, or Configurators.** A good example of shopping guide is wine.com, which acts as an intermediary between hundreds of small wineries around the country and the end customer, who often knows little about shopping for wine. By providing wine suggestions from wine.com's expert tasters, menus, and other gift selections, this Website adds efficiency value to the shopper, who otherwise might need to visit several locations to obtain similar results. Other examples are:

1. The Schwab IRA Analyzer, which aids individual investors in determining what type of individual retirement accounts are right for their particular situation.
2. Carpoint (http://carpoint.msn.com/), which offers customers several ways to get and configure a desired car. It also offers a simple, 6-question recommender (referred to as the Shopping Guide for car buyers to help them narrow down their choices).

• **Show related products and accessories.** When a customer has decided to buy a product, do not force the customer to search for accessories and/or related items. For example, if a customer has decided to buy a personal digital assistant (PDA), providing customers an easy way to add a case or extra styli for the PDA in the shopping cart not only reduces the time it would take the customer to find them but also makes it easier for the online store to cross-sell.

800.com provides an effective design solution for helping users find the desired product (see Fig. 12.1). Not only does the page show the product details, customer reviews, accessories and related items, and other information accessible in one click without losing the product information (by using tabs below the basic product information), it also allows customers to find an item with "fewer" and "more" features. This is a very effective approach to get nonexpert customers quickly to an optimal choice.

FIG. 12.1. By using a "More Features" and "Fewer Features" link, 800.com makes it easy for non-expert customers to select a product with an optimal set of features.

Removing Purchase Friction

Depending on which research you read, roughly 25% to 75% of online shoppers abandon their shopping carts before consummating the purchase. The online stores that fall into this category have failed to follow the fundamental rule of profitable business: Make it easy for people to give you their money. A recent study conducted by Vividence (2001) found the following reasons for shopping cart abandonment (with percentage of participants in parentheses):

1. High shipping prices (72%).
2. Comparison shopping or browsing (61%).
3. Changed mind (56%).
4. Saving items for later purchases (51%).
5. Total cost of items too high (43%).
6. Checkout process too long (41%).
7. Checkout requires too much personal information (35%).
8. Site requires registration before purchase (34%).
9. Site is unstable or unreliable (31%).
10. Checkout process is confusing (27%).

Knowing the reasons for shopping cart abandonment, it's important that the following guidelines be followed to reduce purchase friction.

Disclose All Costs Up-front

Customers are often uncomfortable filling out forms and supplying credit card numbers before they know the total cost of their purchase (including shipping charges and tax). The cost of the transaction is a critical piece of information, so give your customers the grand total up front, with minimal data entry—a zip code for example. Again, be clear about why you're asking for their information. It is important that all the disclosures are made up-front and in simple understandable language. The disclosures are needed for the following:

* Pricing information (and return policy).
* Availability and shipping information.
* Any additional hidden costs.

Make Registration Forms Short and Optional

Some sites require their customers to register before allowing them to place an order. This is one friction to avoid when a customer is ready to complete the sale and give you money. As Chaparro, Childs, Praheswari, and Rappard (2001) observed in their study: "Costco.com received many negative comments and was reported to be 'very complicated.' This was primarily due to the fact that the site required the participant to enter personal information (such as name, address, credit card info) just to put an item into the shopping cart."

If you'd like your customers to register with your site, make it clear that it's their choice but also give them a reason to fill out the form. The reason could be either monetary or nonmonetary benefit to the customer that will encourage customers to part with their personal information. Preferably, during the transaction, ask only for the information essential for completing the transaction, usually the billing and shipping information. If you must gather additional personal information, wait until the transaction is complete and then start gathering secondary information, usually demographic and lifestyle data (Gordon, 2000). The more information the user needs to supply before the purchase, the greater the likelihood that errors will occur or that customers will become frustrated and leave the site.

Make the Shopping Cart Easy to Find

Once users have located the items they want, it is important that customers be able to begin the ordering process. Therefore, the shopping cart should be easy to find and located consistently throughout the site. Amazon.com does this well by putting the shopping cart icon in the top right corner of every page. There should also be multiple entry points into the shopping cart from the site so users can see what is in their cart and add additional items (Bidigare, 2000).

Allow Saving of Shopping Carts

Provide for items in the shopping cart to be saved for future purchases. Customers may not follow through with the purchase for several reasons; for example, they may go to another site to comparison shop or may want to wait to accumulate several items in the shopping cart to reduce their shipping charges (Vividence, 2001). Sites may also want to support similar features commonly referred to as wish lists, gift registries, or shopping lists. This type of feature allows a user to gather items together for a purchase at a future date or create a list to share as gift ideas with friends and family (Bidigare, 2000; see Fig. 12.1 and Fig. 12.2 for examples of how 800.com and Amazon.com lets customers add an item either to the shopping cart or to wish list).

Fast-Track the Checkout Process

Sites should make it more convenient to buy standard or repeat-purchase items (such as the one-click-to-purchase approach at Amazon.com and at 1800flowers.com). The easiest purchase process many of us have witnessed is Amazon.com's "1-Click" purchase. It automatically uses the same credit card and delivery address you used last time and places the order. Here are some additional ways to "fast-track" the checkout process:

• Don't force a "format" for entering credit card information on users. Many online stores place seemingly arbitrary constraints on how information must be

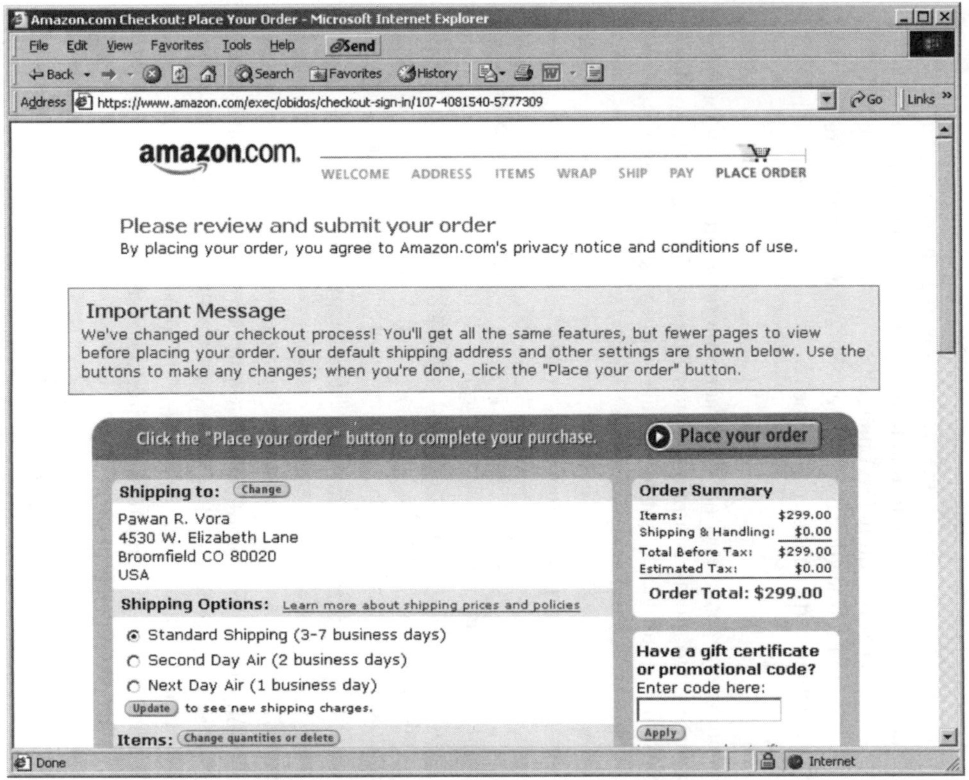

FIG. 12.2. This redesigned page by Amazon.com shows one page that combines all the information: Order Summary, Shipping Information, Shipping Options, Shopping Cart Items, and an option to enter Gift Certificate and Promotional Code. If users don't need to update any information, they could simply click on the "Place Order" button and the order is complete.

entered, from the layout of a phone number to the characters used for dates or entering the credit card number without spaces or dashes.

• Keeping a record of credit card information—of course, with the customer's permission—so that consumers don't have to enter the information repeatedly.

• Offering a variety of payment solutions, which include credit cards, cash cards, payment services (e.g., CyberCash and PaymentNet), payment intermediaries (e.g., PayPal and iEscrow), and electronic checks.

• Also make sure that removing the items from the shopping cart is easy. Chaparro et al. (2001) found that to delete an item from the shopping cart at BestBuy.com, the user was required to change the quantity to 0 and then click on an "Update" button further down the page. Not knowing how to remove an item from the shopping cart could be one of the reasons for customers to abandon shopping carts.

Amazon.com implements a good fast-track checkout process by offering 1-Click purchase option to all returning customers (see Fig. 12.3). Even those returning customers who decide not to use the option, the checkout process is streamlined to just one confirmation step (see Fig. 12.2).

Eliminate Distractions During Checkout

When customers have begun the checkout process, eliminate all potential distractions. Don't provide any links that are not relevant, for example, related products, links to other product categories, banner ads, and so forth. Gordon (2000) refers to it as taking the users into the *tunnel*, which is a clearly defined path that leads the customers through the checkout process and ends with a confirmed sale. Some good examples are Amazon.com and BarnesandNoble.com.

Establish Trust

Like most nonpersonal modes of interaction, developing trust is critical for e-commerce sites as lack of trust has been identified as an important barrier for purchasing online (Cheskin Research, 2000; Cheskin Research and Studio Archetype/Sapient, 1999). Building trust requires both creating a trusting environment and ensuring privacy of a customer's personal information.

Creating a Trusting Environment. A 1999 joint study by Cheskin Research and Studio Archetype/Sapient identified the following factors as crucial in communicating the trustworthiness of a site to its customers: brand, navigation, fulfillment, presentation, up-to-date technology, and logos of security-guaranteeing firms (see Lee et al., 2000; Cheskin Research, 2000).

Prominently displaying logos of security-guaranteeing firms (or trust seals) and using secure technologies are particularly useful in assuaging customer fears

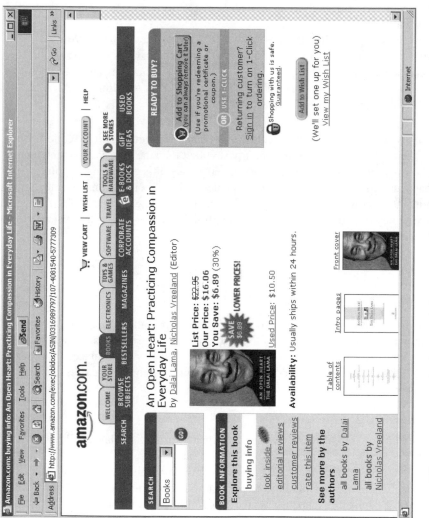

FIG. 12.3. A good example of building trust with customers with the text: "Shopping with us is safe. Guaranteed." This example also shows how returning customers can use 1-Click purchase to make the checkout process quicker.

by emphasizing that their personal information is treated with care. The seals of approval of some private institutions (e.g., Visa and TRUSTe), including banks and privacy organizations, significantly reduce the sense of financial risk, especially for newcomers to e-commerce. Credit card companies are at the top of this list for those less comfortable with Web commerce. Websites should also consider clearly displaying credit card symbols on their home pages to enhance the communication of trustworthiness.

In addition to providing secure Web servers and processes, it may help to provide a satisfaction guarantee (Bidigare, 2000). For example, just below Amazon.com's button to add an item to shopping cart there is bold text that indicates: "Shopping with us is 100% safe. Guaranteed" (see Fig. 12.3). This text links to Amazon's guarantee and frequently asked questions about safety.

Ensuring Privacy. Privacy concerns refer to customers' ability to control the use of their personal information for purposes other than the transactions for which the information was collected. For e-commerce sites, it is extremely important to appease customers' concern that the companies will sell their personal information without their knowledge or permission (Hoffman, Novak, & Peralta, 1999). Thus, it is extremely important that the sites make a full disclosure of how the customers' personal information will be used.

Removing Postpurchase Friction

Keep the Customer Informed

As discussed, it is important that customer expectations are set correctly throughout the purchasing process to avoid unnecessary surprises. It is also important that customers are kept informed as soon as they have placed an order and until the product is delivered. Here are some ways e-commerce sites can keep customers informed:

- Sending e-mail for order confirmations.
- Sending an e-mail when an order is shipped (or if it is delayed). Preferably, the order should ship on-time or before-hand.
- Allow users to track the status of their order online. This may be accomplished via offering appropriate links to the online locator functions offered by the package delivery companies such as Federal Express, United Parcel Service, and the U.S. Postal Service.

Fulfillment Rules!

A must: Deliver on-time or before-hand. According to Binary Compass Enterprises, on-time delivery is the most important factor in purchase decisions.

No doubt, there will be situations when fulfillment promises won't be kept. In such cases, consider offering free merchandise or discounts on future purchases.

When You Say Customer Service, Mean It

The customer experience does not end when the purchase is made. For both bricks-and-mortar stores and online stores, customers expect a level of quality and support that must be delivered on time. To meet customers' expectations:

- Provide 24/7 (24 hours a day, 7 days a week) support, which is becoming imperative in today's business climate.
- Respond to customer e-mails in a timely manner. In my experience, customers expect a response to an e-mail inquiry within 24 hours. If responding within 24 hours is not possible, it is better to set up reasonable expectation for the response time so as to minimize a customer's frustration.
- Offer access to frequently asked questions (FAQs) that provide users with additional information without the need to contact a customer service representative is particularly useful. Lohse and Spiller (1998b) reported that featuring a frequently asked questions section in an online store was associated with more traffic.
- Provide a searchable and browsable product and service knowledgebase (e.g., Microsoft's Knowledgebase, http://search.support.microsoft.com/kb/; Dell's Personalized Support, http://support.dell.com/).
- Link customers to "expert" sites that address some of the support questions or providing moderated customer discussion groups so that customers can help each other.
- Provide preaddressed return packaging to make returns easier.
- Do not hide customer service contact information. List a toll-free phone number and offer the callback option. And consider offering live chat with customer support personnel and guidance from experts.

The challenge for reducing postpurchase customer service-related friction is not only to choose appropriate service methods but also to continuously enhance the selected methods as customer demands and expectations change.

Let Customers Control Their Information

It is important to provide customers with complete access to their information. Specifically:

- Allow customers to change their personal information.
- Allow customers to opt out of the e-mail newsletter or any other form of communication.

- Allow customers to opt out from allowing you to aggregate and resell or otherwise manipulate and use their purchasing patterns and history.

DISCUSSION

In this chapter, an attempt is being made to capitalize on the value proposition associated with Internet commerce—that is, providing friction-free interaction with the customer for the processes of finding, ordering, and receiving it. It's important to ensure that making a purchase on an e-commerce site is as easy as possible and without any unnecessary distractions. If people are coming to your site to buy, make sure you don't distract them from that primary intention. Providing that extra little bit extra is always helpful, but if it begins to interfere with the simple task of placing and paying for an order, you are likely to annoy your customers—especially those who use your site often. Remember: Improved customer experience does not mean giving preference to "wowing" the customer over making it possible for customers to achieve their goals easily and quickly.

Every substandard interaction that is introduced to a customer within a store is a barrier that lies between the customer and a completed transaction. Therefore, it's worth spending the extra time, money, and effort required to remove the barriers and nuisances from your purchasing process. Looking at the failures of many innovative e-commerce sites provides substantial credence to the notion of designing friction-free experiences. Rosenberg (2000) articulates the issue very eloquently:

WebHouse[1] faced a kind of frictional resistance that a lot of e-commerce schemes are encountering—the friction generated when complex innovations rub against entrenched consumer habit . . .

But you also have to make sure that potential customers can quickly grasp what you're doing and why they should bother learning something new. That's where so many e-commerce ventures—whether group-buying deals or price-comparison shopping bots or reverse auctions—falter. How many shoppers have the time or patience to figure out these unconventional schemes and make sure they are in fact to their benefit? And even if we're satisfied on that score, how many of us feel that the few cents or dollars they promise to save us are worth their precious time? . . .

[e]Bay is an example of a company that created a new marketplace online, reduced the general friction level by putting masses of individual buyers and sellers in touch with one another, enabled a huge volume of new transactions and managed to grab a small slice of them for itself.

But for every success story like eBay there are probably a hundred flops tanking—because they failed either to make things easy for the consumer, to account for the "frictional" resistance of old-economy competitors or to build profits into their plans.

---

[1] The WebHouse concept was classic friction-free thinking. The idea was to apply the Priceline.com model, in which customers bid for unused airline-seat inventory, to gasoline and groceries.

Applicability of the Concept of Friction-free Experiences

This chapter was focused on providing guidelines for designing friction-free experiences for business-to-consumer e-commerce sites. The same principle of designing friction-free experiences can be applied to other e-commerce sites as well, including those focusing on business-to-business applications. When designing experiences for those sites, we need to remember the following:

- Focus on helping users achieve their goals quickly and not getting lost on the tasks or activities and take advantage of the nonlinearity of the hypertext-like environment afforded by the Web.
- Think in terms of repeat visits of the customers.

REFERENCES

Bellman, S., Lohse, G., & Johnson, E. J. (1999). Predictors of online buying behavior. *Communications of the ACM, 42*(12), 32–38.

Bhatnagar, A., Misra, S., & Raghav Rao, H. (2000). On risk, convenience, and Internet shopping behavior. *Communications of the ACM, 43*(11), 98–105.

Bidigare, S. (2000). *Information architecture of the shopping cart. Best practices for the information architectures of e-commerce ordering systems.* [online: a White Paper by Argus Associates.] Available: http://argus-acia.com/

Chaparro, B., Childs, S., Praheswari, Y., & Rappard, A. (2001, Winter). Online shopping: TV's, Toasters, and Toys! Oh My! *Usability News* [online]. Available: http://psychology.wichita.edu/surl/usabilitynews/3W/web_usability.htm

Cheskin Research. (2000). *Trust in the wired Americas* [online]. Available: http://www.cheskin.com/

Cheskin Research and Studio Archetype/Sapient. (1999). *ECommerce Trust Study* [online]. Available: http://www.sapient.com/cheskin/assets/images/etrust.pdf

Cooper, A. (2001). Navigating isn't fun [online]. Available: http://www.cooper.com/newsletters/2001_10/navigating_isnt_fun.htm

Cox, D. F., & Rich, S. (1964). Perceived risk and consumer decision making. *Journal of Marketing Research, 1*, 32–39.

Gordon, S. (2000). Is your site shoppable? Convert browsers to buyers [online]. Available: http://builder.cnet.com/webbuilding/pages/Business/Shoppable/ss01.html

Greenwood, W. (2001). Beating the checkout blues. http://www.cooper.com/newsletters/2001_06/beating_the_checkout_blues.htm

Hammer, M. (2001). *The agenda. What every business must do to dominate the decade.* New York: Crown Business.

Hoffman, D. L., Novak, T., & Peralta, M. (1999). Building consumer trust online. *Communications of the ACM, 42*(4), 80–85.

Hoque, A. Y., & Lohse, G. L. (1998). An information search cost perspective for designing interfaces for electronic commerce [online]. Available: http://citeseer.nj.nec.com/hoque98information.html

Hurst, M. (1999). *Holiday '99 e-commerce: Building the $6 billion customer experience gap* [online]. Available: http://www.creativegood.com/holiday99/

Kalakota, R., & Whinston, A. B. (1996). *Electronic Commerce: A Manager's Guide.* Boston: Addison-Wesley.

Jupiter Media Metrix. (2001). *Holiday 2001 online shopping results: Traffic up 50 percent compared with last year* [online]. Available: http://www.jmm.com/xp/jmm/press/2002/pr_010702.xml

Lee, J., Kim, J., & Moon, J. Y. (2000). What makes Internet users visit cyber stores again? Key design factors for customer loyalty. In CHI '00 Conference proceedings (pp. 305–312). California: ACM Press.

Lohse, G. J., & Spiller, P. (1998b). Quantifying the effect of user interface design features on cyberstore traffic and sales. In *CHI '98 Conference proceedings* (pp. 211–218). California: ACM Press.

Lohse, G. L., & Spiller P. (1998a). Electronic shopping. *Communications of the ACM, 41*(7), 81–87.

Riggins, F. J. (1999). A framework for identifying Web-based electronic commerce opportunities. *Journal of Organizational Computing and Electronic Commerce, 9*(4), 297–310.

Rosenberg, S. (2000). Keep the customer dissatisfied [online]. Available: http://www.salon.com/tech/col/rose/2000/10/20/customer/index.html

Spool, J. (1999). Designing Websites for business. *Eye for Design, 6*(2), 4–7.

User Interface Engineering. (2001b). *Users continue after category links* [online]. Available: http://world.std.com/~uieweb/Articles/continue_after_categories.htm

User Interface Engineering. (2001a). *What causes customers to buy on impulse* [online: E-commerce White Paper]. Available: http://www.uie.com/

Vividence. (2001). *Holiday readiness 2001: Helping retailers prepare for the holiday season* [online]. Available: http://www.vividence.com/

13

Web Support for Context-Sensitive Interaction

Chris Johnson
Yun-Maw Cheng
University of Glasgow, Department of Computing

INTRODUCTION

Many claims have been made about the potential benefits of mobile, context-sensitive computing. The initial focus was on "follow-me" applications in which the presentation and dissemination of information was tailored to the users' tasks as they moved from one working environment to another. More recently, attention has shifted from systems that sense the location of users toward architectures that help users sense systems and other objects in their environment. At the same time, we have seen the increasing availability of personal digital assistants (PDAs) with Internet access, for instance, through the 802.11 protocol or through cellular telephone networks. One consequence of this is that the Web has become an attractive infrastructure for the provision of context-sensitive services using a broad range of sensing technologies. The Glasgow Context Server (GCS) is intended to help interface designers validate the claimed benefits of location-sensing, location-disclosing and environment-sensing applications.

Researchers in the field of human–computer interaction have been interested in location sensing as a means of supporting context awareness for a considerable period of time. The active badge system was conceived in the early 1990s and used a network of infrared "base sensors" to locate the badges that users wore as they moved around a building (Harter & Hopper, 1994). The subsequent PARCTab system extended this initial idea; devices that resemble pagers replaced the badges. These provided location information to the system and also offered their users a limited range of services as they moved within a building (Want et al., 1995). More recently, the focus has moved from systems that sense the location of their users toward architectures that help users sense systems and other objects in their environment. This work has drawn on attempts to integrate physical and digital space (Ishii & Ullmer, 1997; Want, Fishkin, Gujar, & Harrison, 1999).

A TAXONOMY OF CONTEXT-SENSITIVE INTERACTION

One consequence of recent developments has been considerable disagreement over terms such as *context awareness* in human–computer interaction. Previous generations of location-aware systems claimed a form of context sensitivity because they used information about the position of the user to make inferences about their likely activities. This enabled them to filter information and to offer input options that were intended to support users' predicted tasks. The recent change in emphasis toward systems that actively sense the objects in their environment has led a number of authors to distinguish between these systems and those that simply make inferences based on location information (Cheng & Johnson, 2001). Context awareness implies the ability to directly sense properties of the user's environment rather than simply making predictions based on the user's probable location. Figure 13.1 illustrates the distinctions previously mentioned and also provides a high-level overview of the architectures that will be described in the remainder of this chapter.

This diagram does not use the term *awareness*. This word can have the misleading effect of encouraging strong claims about the ability of these systems to consistently provide their users with appropriate information. Current technology supports a very primitive form of "awareness." All existing systems make frequent mistakes about users' tasks and information requirements. It should be noted that Fig. 13.1 is not intended to provide a complete overview. Further distinctions are likely to emerge with new forms of context-sensitive interaction.

The domain of context-sensitive systems can be divided into three types of interactive applications. The first includes systems that disclose information about the user's location, such as the Active badge systems previously mentioned (Harter & Hopper, 1994). This information can inform subsequent interaction. For instance, a user might prefer to telephone rather than e-mail a colleague who is out of the building. Active badge systems need not, however, provide any location-dependent information to the person carrying the badge. In contrast, location-filtering systems exploit knowledge of the users' position to tailor the information that is presented to them. Examples include the Guide (Cheverst, 1999) and Cyberguide (Abowd

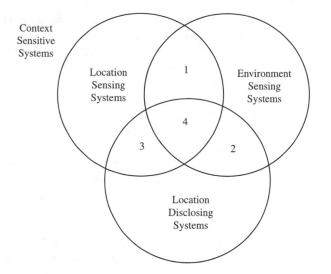

Context
Sensitive
Systems

Location
Sensing
Systems

1

Environment
Sensing
Systems

4

3   2

Location
Disclosing
Systems

FIG. 13.1.   The relationships between different types of context-sensitive systems.

et al., 1997) applications. A third type of context-sensitive system actively senses objects, people, and properties of a user's immediate environment. In their pure form, these sensing systems need not have any explicit data about the user's position. For example, Bluetooth applications can detect and create links between enabled devices without any location information.

Figure 13.1 also illustrates the intersections between these different approaches. For example, Region 1 represents applications that detect the user's location and other objects in their environment. Applications in this region need not, however, disclose information about the user's location to other people. This approach is embodied in Hewlett-Packard's CoolTown proposal (Kindberg et al., 2000). The proponents of this form of "nomadic" computing are concerned to avoid the privacy concerns that have affected previous generations of location-disclosing systems. In contrast, applications in Region 2 disclose information about the user's location and sense information about their environment but do not tailor interaction using knowledge of the user's location. There are few examples of this form of interaction. Most interface designers seem to be eager to exploit information about the user's location and the objects in their environment to inform subsequent interaction (Cheverst, 1999). Systems in Region 3 tailor the presentation of information to their users depending on their probable location. They would also disclose information about that person's location but would not actively sense other objects in their environment. The PARCtab system is an example of such an application (Want et al., 1995). The Glasgow Context Server was designed to provide a flexible test bed for the prototyping and evaluation of context-sensitive human–computer interfaces.

It is important to stress that relatively little is known about the usability problems that might arise from interaction with these different context-sensitive applications. Previous authors have identified privacy concerns associated with the disclosure of location information. Others have identified some of the problems that arise when users have to "repair" the inappropriate predictions that context-sensitive systems make about likely user tasks. These observations are, however, often based on intuition and introspection rather than empirical studies. There are a number of reasons for this. The development of context-sensitive systems requires considerable engineering skills. The teams that have the necessary expertise to implement these applications often focus on innovative system architectures rather than detailed usability studies. Unfortunately, the engineering focus of previous research has done little to reach any consensus about appropriate hardware or software architectures. For example, a bewildering array of middleware has been developed to support context-sensitive interaction (Cheverst, 1999). Some systems exploit commercial systems, such as CORBA. Others rely on bespoke software that often has significant benefits in terms of functionality but which is not widely used by others implementing similar systems. This diversity, in part, reflects the high level of interest in context-sensitive computing. It also poses significant problems for anyone who is more interested in the development and evaluation of novel interaction techniques than they are in the underlying technologies.

A number of recent projects have turned away from more esoteric software and are exploiting Web protocols, in particular the underlying HTTP mechanisms, to support the implementation of context-sensitive systems (Cheng & Johnson, 2001; Kindberg et al., 2000). This has a number of benefits. Web protocols provide well-understood infrastructure for developers. This is essential if research projects are to be propagated beyond the laboratory bench and into end-user environments. The Web also offers a number of benefits for interface designers. Context-sensitive applications can be integrated with recent development in mobile Internet services. As we shall see, however, there are a number of technical problems that remain to be overcome before designers can use the Web to support the prototyping of context-sensitive interaction.

USING THE WEB FOR LOCATION-SENSING SYSTEMS

The GCS is designed to provide a low-cost test bed for the use of indoor location detection in human–computer interfaces. The intention was to provide a means of exploring whether or not some of the claimed benefits for this form of interaction could actually be realized by the evaluation of prototype systems. The emphasis on low-cost techniques was born from pragmatic constraints. We had limited software development resources and wanted to entirely avoid any reliance on esoteric hardware. We also wanted to avoid any changes to the infrastructure or fabric of

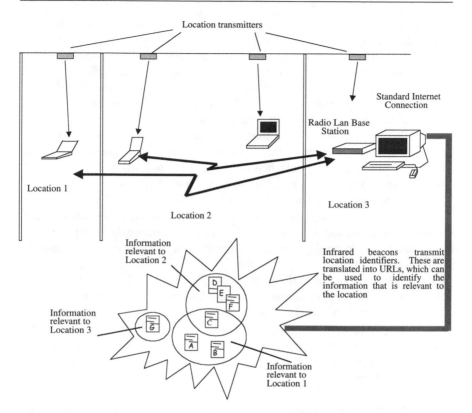

FIG. 13.2. The GCS architecture for location sensing.

the buildings that would house the system. GCS applications should be simple to maintain and should not require periodic recalibration in order to continue to provide accurate location information. A further requirement was that the GCS architecture should be "future proof"; it must be possible to update system components when new technologies, such as Bluetooth or new Web protocols such as HTTP-NG, become widely available. Figure 13.2 illustrates the initial architecture of the GCS system. Our early work focussed on passive location-sensing systems. GCS uses low-powered, diffuse infrared transmitters that send out a unique location identifier. These devices do not actively sense the presence of other transmitters.

As mentioned, we are eager to exploit "off-the-shelf" components. Therefore, we use domestic remote controls from television and audio systems to generate location identifiers. Unfortunately, the protocols used by the built-in IR transceivers for most mobile devices are different from the physical and data link layers used by domestic remote controls. Consumer devices exploit pulse, space, or shift encoding for infrared transmissions (DuBois, 1991). They are different from the Infrared Data Association (IrDA) protocol that is employed by infrared transceivers on most mobile devices. These technical considerations might seem to have little to do with usability issues. Unfortunately, these details have a profound impact on the usability of any eventual system. IrDA transmissions require relatively careful aiming. They rely on directed transceivers that are deliberately intended to support line-of-sight communication. The transfer of information will fail if one of the devices is moved. To exploit this protocol, users would have to carefully line up their device with a transmitter. If any other object came between the mobile device and the transmitter, then the signal would fail. Therefore, we developed software that temporarily replaces these protocols with the "raw infrared mode." Programmers can then gain to access the infrared port as a normal serial port with an infrared transceiver attached. This supports the robust transmission of infrared location information and does not force users to carefully position their PDA within the line of sight to an IrDA device. It should be emphasized that the GCS has a modular design that is consistent with both the Bluetooth standard and with differential radio signal detection. These more elegant approaches can simply be substituted for the infrared beacons that we currently use as a pragmatic alternative.

The GCS uses a commercial wireless local area network (LAN) to support the transmission of location information from mobile devices. However, even with maximal coverage, there will be areas in which connection cannot be guaranteed. This raises a vast range of usability issues (Ebling, John, & Satyanarayanan, 1998). The lack of uniform radio coverage led us to use low-cost infrared transmitters to index into precached information on each PDA. The cache is then periodically refreshed when the user moves into radio coverage. A natural extension of this approach is to do away with the radio LAN entirely and simply use the infrared transmitters to index into a cache of Web pages if the application information is of a relatively small size and if no interaction is required with remote resources. In contrast, Fig. 13.3 shows an example of the distribution of radio transmitters and infrared beacons in the GCS architecture. A user can walk from radio coverage in cell 1 through an uncovered area to reach radio cell 2. On their way, they pass by infrared beacons D, B, and A. For a user to access information about D, B, and A, it must be preloaded before the user leaves cell 1. In other words, GCS must anticipate the user's likely trajectory through the uncovered region. This trajectory is at the heart of the caching mechanisms that support the development of interactive, context aware systems using the GCS.

As mentioned, the GCS was initially conceived as a location-aware system that could be built from off-the-shelf components. We did not explicitly start out with the objective of linking context sensitivity to a Web-based model of interaction. It soon became apparent, however, that the Web offered considerable support for the development of mobile interactive systems using the GCS architecture. There are a number of reasons for this. Web technology runs on a heterogeneous collection of devices. Both servers and browsers have been developed for many different hardware platforms and operating systems. They are also being integrated into a small but growing range of consumer appliances. This development supports some

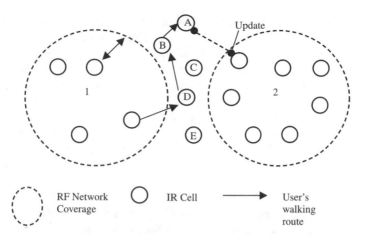

FIG. 13.3. Network topology for offline support.

of the more visionary scenarios that have been developed to support the notion of nomadic computing. Therefore, any context-sensitive system that exploits Web technology, avoids many of the overheads associated with supporting an increasing diverse range of devices:

> In recent years, there has been a proliferation of types of device and access mechanisms using the Web, extending far beyond the conventional personal computer. These access mechanisms range from Web tablets, appliances and TVs in the home, to mobile devices including phones and PDAs, and access mechanisms for the physically challenged. Connectivity capabilities have also evolved to include high-bandwidth modems, LANs and wireless networks... Users now expect to get to critical information through different access mechanisms from different locations and at different times during their day.... The key challenge facing (web content authors) is to enable their content or applications to be delivered through a variety of access mechanisms with a minimum of effort. (Gimson, Finkelstein, Maes, & Suryanarayana, 2001)

There are further benefits from exploiting the Web as component of context-sensitive systems. The relative simplicity of the HTTP protocol and associated addressing schemes has fueled the rapid expansion of Web access. Kindberg and Barton (2001) also argue that the Web provides "just enough middleware." The infrastructure does not assume that devices will all run middleware, such as Java or CORBA, which impose heavy burdens on any application. Similarly, the Web does not assume the presence of global services. It is possible to access local servers if wider connectivity is denied through accident or design. This supports the use of location beacons to index into local caches in the manner described by Fig. 13.3.

A group of pages can be associated with each of the locations that are represented by the infrared transmitters. Instead of expecting the user to explicitly request the pages associated with each of these locations, the GCS location aware system can automatically present them to the user once the infrared signal has been detected. If the user is within radio coverage, inside cell 1 or 2, then the GCS can automatically request any relevant pages that are not already cached on the mobile device from a Web server. If the user follows a trajectory that will take them outside of radio coverage, then it is also possible to cache any pages that the designer predicts might be accessed before the user returns to a radio cell. The Web infrastructure offers considerable support for the implementation of this approach. Typically, servers will provide clients with HTTP headers that specify a cache time-out for each page that a user might request. The header can also indicate when a document was last changed.

Context-sensitive systems can exploit these attributes of the HTTP in a number of ways. For example, if a user has cached a large document and he or she is about to leave a radio cell, then the browser need only check the header to determine whether or not the user will need to download a more recent version of the document. If the current version is up to date, then the user need not waste critical communications resources in initiating a new download. Similarly, if the cache time-out is greater than the disconnection period, then the system need not refresh the page when the user returns to radio coverage. Conversely, servers can use the HTTP header to specify that certain pages should not be cached at all. This is useful for highly dynamic information, such as the location of adversaries in the game that is described at the end of this chapter.

Unfortunately, the task of compiling a cache for offline interaction is complicated. It is difficult to identify all of the potential pages of information that the user might require. Similarly, if location information is used to index into a cache of previous collected Web pages, there is typically no easy means of determining how long to hold those pages or when to update them as the user moves from one location to another. The traditional model of Web-based interaction is almost entirely driven by the user's explicit actions rather than by inferences about his or her likely needs in a particular location. In some cases, it is possible to use knowledge of the layout of a building to inform the selection of pages for presentation outside of radio cells. For example, if a long corridor connects two areas of coverage, then designers can make strong predictions about the user's likely route and hence his or her information needs. Alternatively, users can be given directions that explicitly guide their movements. Designers can then predict the information that the system needs to cache as users follow a prescribed route between radio cells (Davies, Cheverst, Mitchell, & Friday, 1999).

If the user moves away from an anticipated path, then a location-sensing system may not be able to provide access to any associated Web pages that have not been cached before leaving radio coverage. The system can, however, continue to store information about the user's route. The system can then offer the user the option of viewing any missing information that is associated with the infrared cells that they have visited once they regain radio coverage. Designers can also exploit this route information to improve the cache content for future users of the system.

The existing Web infrastructure supports these techniques; however, a number of detailed changes must be implemented. In connected operation, most browsers report a "146, connection refused" warning if they cannot access a server. In disconnected mode, the GCS also logs this information to improve cache performance. Further changes must also be made if standard browsers are to support location sensing. For instance, HTTP requests must be automatically generated if the GCS detects an infrared cell. Requests to fill the cache must also be triggered if the user's trajectory will lead them out of a radio cell. Conversely, user input, including requests for Web pages that are not in the cache, must be flushed from the cache when they return to radio coverage.

The designers' task of supporting such forms of interaction is complicated by a number of pathological situations in which a user might repeatedly move in and out of radio coverage. This can place a heavy demand on network resources as the system attempts to update the cache with appropriate Web pages. Fortunately, the overheads associated with such movements can be reduced by only retrieving documents whose cache life has expired. This can be determined using the HEAD component of the HTTP, mentioned in previous paragraphs. Again, it is important to emphasize that such detailed considerations can have a profound impact on the usability of any location-sensing system that exploits Web technology. If these techniques are ignored, there is a danger that the performance of any implementation will be too slow to support the evaluation of context-sensitive modes of interaction. This example also illustrates the need to develop appropriate prototyping environments that might enable designers to explore the human factors issues of context-sensitive interaction without having to face such relatively low-level issues.

USING THE WEB FOR
LOCATION-DISCLOSING SYSTEMS

Figure 13.4 shows how the initial GCS architecture must be adapted to support the development of location disclosing, or active badge, systems.

As can be seen, the underlying hardware infrastructure remains unchanged. All of the redesign activities are focused on the Web-based infrastructure. In the initial version of the system, changes in the user's location were detected by the infrared transceiver on their device. These were translated into HTTP requests that were predetermined by the designer. In contrast, a location disclosing version of the GCS system sends update information to the server every time a new infrared cell is detected. CGI scripts can then be used to update Web pages that disclose the location of the user's machine. Such changes are relatively straightforward. It is also possible to create hybrid systems that combine elements of this approach with the presentation of location-dependent information that was described in

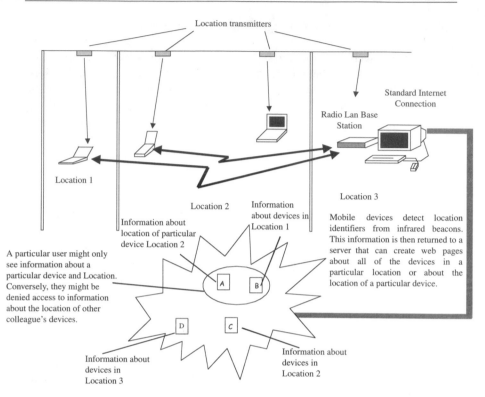

FIG. 13.4.   The GCS architecture for location-disclosing interfaces.

previous paragraphs. This creates interfaces that are characterized by Zone 3 in Fig. 13.1.

A number of usability issues stem from our use of the Web to disclose information about a users location. First, because we are using low-cost passive sensors, a user's location will only be updated if he or she has a PDA or other mobile device switched on. One consequence of this is that the location data will not be as accurate as "true" active badge systems that continually update the system of the user's location at all times. We are, however, prepared to accept this limitation given the privacy concerns that have been expressed about these more active forms of sensing.

A second usability concern stems from our use of the Web as a platform for location disclosure. In the simplest case, a Web page that provides information about the user's location could be retrieved by anyone that has access to the host server. We were therefore concerned to introduce access control mechanisms so that a user could explicitly grant their colleagues the right to view this information. There are three different means of restricting access to Web pages using standard browser technology. It is possible to force users to provide a password before they can access particular pages of information. It is also possible to restrict access to a server using the client's Internet address. Finally, documents can be protected using public key cryptography. Both the request for the document and the document itself are encrypted so that the text can only be read by the intended recipient(s). Public key cryptography also enables the client to authenticate the source of the information. Initial versions of the GCS architecture employed password protection to enforce access control. This introduced significant overheads. It proved to be a nontrivial task to set up and administer the privileges that were associated with particular users. This approach is also relatively inflexible. It is difficult to temporarily suspend access privileges if a user wanted to prevent disclosure from a subset of their colleagues. Encryption provides a more flexible approach in which temporary changes can be made to the keys that are needed to access particular items of information. This simplifies the implementation of fine-grained access control policies because protection mechanisms are associated with resources.

The converse problem is, of course, that it can be difficult for designers to enumerate all of the potential users who might have the key that is necessary to access an encrypted page of information. Finally, it is possible to restrict access to users associated with a particular Internet address. This offers significant advantages in terms of the "usability" of any context-sensitive system because users are not explicitly prompted to remember passwords and keys. Unfortunately, significant problems affect the two standard means of implementing this approach. Server access restrictions can be placed at the directory level. This is a flexible approach that can be implemented by information providers. However, it can be error prone when individuals forget to remove temporary restrictions or to apply necessary prohibitions. Server access restrictions can also be applied at a global level. This has the benefit of creating clear policies for users to follow. Unfortunately, it lacks the flexibility of more fine-grained approaches. It should also be noted that a number of companies, including Microsoft and Netscape, have launched browser extensions and associated services that provide significant support for the maintenance of access control policies over the Web. This area is also a focus for recent work on the future infrastructure of the Web (Stein & Stewart, 2001a).

The higher level point is that the lack of fine-grained, Web-based access control mechanisms can have a profound impact on users' confidence in location-disclosing systems. Many of the users who participated in the initial design of the GCS system expected to implement a broad range of access control policies. Some individuals wanted to grant their colleagues the right to view their locations wherever they went. This led to conflict because other users wanted to deny access to information about people who were gathered in particular locations. For example, the head of one of the organizations that we studied was reluctant to transmit information about workers who might be in his office on confidential or personal matters. Such observations illustrate the way in which the detailed implementation of the underlying Web-based infrastructure had to be tailored in response to the observations that emerged from initial usability trials. Access control should not only protect the privacy of particular users. It may also have to be applied to all users in a particular location.

USING THE WEB FOR
ENVIRONMENT SENSING

A number of research groups have identified the potential benefits that might be gained if users are provided with more active, environment-sensing systems. These applications go beyond the location-sensing and location-disclosing approaches that have been described in previous paragraphs. In contrast, this new generation of active systems form part of a wider vision in which users are increasingly nomadic but will require continuous access to a range of computational resources "at work, at home, at play." There is also an expectation that they will be able to access these resources in a quick and convenient manner. Users must be able to discover where the nearest resources are to be found. They must then be able to negotiate access to printers, the Internet, and a broad range of devices in many different physical locations. This work is particularly relevant within the context of this book because some of these proposals are explicitly based on Web technology. For instance, Hewlett-Packard's CoolTown project associates URLs with objects in the user's environment. Once a mobile device can access these identifiers, they can download and display the pages of information that are associated with these objects:

> As you enter a CoolTown conference room you can collect the URL for the room from a beacon or tag into your PDA or cell phone. The URL will lead to a Web page for the room giving links to, for example, the room's projector, printer, and electronic whiteboard, as well links to Web-based maps . . . Workers in the room can also "eSquirt" URLs at the room's "Web appliances" such as Web-enabled projectors or printers. An eSquirt is just a wireless transfer of a URL over a short-range wireless link. Squirting a URL at the projector will project the corresponding Web page, creating a shared Web browser. Squirting a URL at the printer will fetch the Web pages and print them." (Kindberg et al., 2000)

This quotation illustrates how environment-sensing systems can be built from the same technologies that the GCS has used to support both location-disclosing and location-sensing systems. Short-range signals provide the identifiers, or the CoolTown URLs, of the objects and people in the immediate vicinity. The information that is associated with these addresses can then be retrieved over longer range networks.

Figure 13.5 illustrates the way in which the GCS architecture can implement a variant of the CoolTown approach to environment sensing. Instead of associating infrared transmitters with fixed locations, they can be attached to objects including digital projectors, printers and mobile devices. The transceivers on a user's PDA

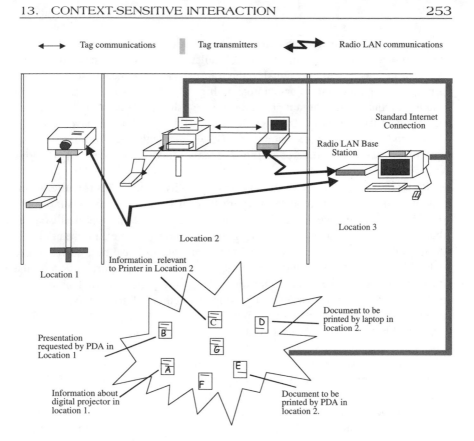

FIG. 13.5.   The GCS architecture for environment-sensing systems.

can then detect the signals that are generated by each of these objects. There are two means of accessing the information that is associated with these infrared signals. The proponents of the CoolTown project envisage the implementation of a protocol whereby these short-range signals are directly translated into URLs. This is not possible in the GCS architecture. We have opted to use off-the-shelf domestic remote controllers to reduce the costs and complexity of implementing the system. These can be programmed to generate relatively simple numeric patterns that cannot easily be associated with the components of a URL. In consequence, we rely on a "lookup table" that associates a URL with each of the different patterns programmed into the remote controls. When the user's device detects a new infrared signal, it uses the table to look up an associated URL. If the pattern is not found, then the device can communicate with a remote server over the radio LAN to retrieve either the URL, which can be stored for later reference, or the page that is associated with the URL and the infrared pattern.

The eSquirt applications that are a strong feature of Hewlett-Packard's Cool-Town system can be implemented in a similar fashion. In this case, the user's device transmits an infrared signal to the transceiver on another device, such as a printer or digital projector. For example, in Fig. 13.5 the PDA in Location 1 might eSquirt a request to the digital projector to begin showing a presentation that can be retrieved from a particular URL. Unfortunately, we did not have access to the Internet-enabled devices that are often envisaged within environment-sensing systems. We were able to overcome this problem by linking these devices to laptops with radio LAN access.

Problems arose because we are relying on "rawIR" mode. This is entirely consistent with the ideas behind the eSquirt style of interaction. Users simply fire relatively brief messages at devices to index or reference the remote resources that are to be used; they do not engage in the more complex information exchanges that are associated with the IrDA protocol. We have not, however, developed a protocol that would enable the receiver of the eSquirt to decode the infrared message and extract an arbitrary URL for a presentation. This remains the subject of ongoing research. The implementation of the protocol is relatively trivial. It is far harder to predict the range of commands that should be supported between the user's PDA and devices such as Internet-enabled digital projectors.

A number of further limitations frustrate attempts to use the Web as a principal component in environment-sensing systems. As we have seen, HTTP does not support the fine-grained access control that must be implemented to protect users' location information. In environment-sensing applications, similar support is required to ensure that nomadic users have the relevant permissions to access the resources that they discover on their travels. Similarly, it is often necessary to exclude users from devices that are being used by their colleagues. Otherwise, a member of the audience could simply eSquirt a request to change the presentation while the speaker was talking.

A further limitation of the Web-based architecture is that there is no explicit support for information discovery. The fundamental concept behind these sensing systems is that users should be able to query their environment to find out what resources it might offer. Previous paragraphs have described how lookup tables can be used to link infrared identifiers with URLs. These can either be cached or can be held on a central server. Neither approach is ideal. If mobile devices cache the lookup table, it can be difficult to arrange access to any devices that users might find in an unexpected location. Conversely, holding these tables on a central server can create problems for users during periods of disconnected interaction.

A number of higher level criticisms can also be raised about more ambitious visions nomadic computing. For example, there is a lack of practical evidence to support the claimed benefits for many of the proposed applications. This is a significant issue, given that the mobile Internet has not enjoyed universal success. The perceived failure of WAP services to meet user expectations provides an instructive example. Strong claims about the utility and "usability" of mobile technology should be avoided until there is direct evidence of benefit to their intended users. We are, therefore, in the process of evaluating a number of different GCS applications. The intention is to map out the usability issues that affect a broad range of location-filtering, location-disclosing and environment-sensing systems.

THE VIRTUAL NOTELET CASE STUDY

The GCS architecture is being used to implement a system that enables its users to stick virtual notes to physical objects, such as doors. This application was chosen because a number of authors had claimed significant benefits from such systems (Baldonado, Cousins, Gwizdka, & Paepcke, 2000, Pascoe, 1997). We were anxious to determine whether these benefits could be demonstrated during a series of evaluations. The initial design focussed on mobile users who could retrieve notes by pointing their PDAs at an infrared transmitter. The PDA received an identifier from the transmitter that was then translated into an HTTP request to be transmitted over the radio LAN. The PDA then presented the user with the Web page that was associated with the infrared beacon. The initial design also enabled users to append information to this note through a series of further HTTP and CGI transactions.

Our intention was to augment and not replace the use of physical notes. With this initial model in mind, development progressed by observing the many different ways in which people use the paper-based medium. We soon realized that a computer-based system could not match the flexibility of physical notes. However, we hoped to offer a range of additional functionality that might support the conventional use of these physical annotations. For instance, we designed a note that users could update with their location as they moved around a building. This was based on the location-disclosing architecture illustrated in Fig. 13.4. A Web page was used to publish this location information as the user's PDA transmitted new infrared identifiers to the GCS over the radio LAN. We also designed a note to display the user's agenda for the day. A colleague could retrieve a page of information about their coworker's activities by pointing a PDA at an infrared transmitter that was mounted on their door. The tight coupling between the GCS and Web architectures again eased the implementation of this facility. The agenda was already available from an online time management system.

As mentioned, our initial design focused on providing a notelet service to nomadic users. People could only access virtual annotations if they were standing within range of an associated infrared beacon. This was a deliberate design decision; we were keen to follow the models proposed by the authors previously cited. It was not possible for users in a remote location to determine whether any notes had been left on the door of their offices. During initial usability trials, we quickly discovered the importance of this facility for many users. Such insights emphasize the importance of conducting field studies to validate the many proposed benefits of context-sensitive systems.

Our choice of a Web-based infrastructure eased the implementation of this facility. Users were provided with the URL for the page associated with their office so that they could retrieve any notes that had been left there wherever they could obtain Internet access. Unfortunately, our response to users' requests raised a host of security concerns. The location of a user was potentially exposed to anyone with access to the Web. Therefore, we implemented fine-grained access control mechanisms based on password mechanisms. Previous sections have argued that this is a relatively inflexible approach. It can be difficult to temporarily alter access privileges, for instance, if a confidential meeting is taking place. Fortunately, all users of the GCS can mask their location at any time by switching off their PDAs. In the future, however, it may be possible to benefit from the more elegant access control mechanisms that have been proposed for successors to HTTP.

Unfortunately, the continuing usability studies revealed a host of further problems that were not addressed by the introduction of these relatively simple privacy mechanisms into the Web-based model of interaction. For instance, physical notelets provide their readers with a number of different sources of information. They provide the message that is written on them. They also provide an indication that the person who wrote the message was previously in the same location as the note that they have left. This is important because the recipients of these notes often make a number of important inferences based on this information. For example, they may conclude that the writer is nearby if they have left a note in the last few minutes. Unfortunately, by allowing remote, Web-based access, we not only provided additional facilities to the users of the virtual notelet system, we also prevented people from making these forms of inferences. In pathological cases, the system enabled users to deliberately mask their locations by leaving messages that indicated they were in one position when they were in fact in another.

The evaluation of the notelet system also identified a number of further usability issues that centred on the naming of particular locations. The initial prototype assumed that the uses would have to be within range of an infrared transceiver in order to read the messages that had been left in that location. This built on the idea that users have to be close enough to a physical notelet before they can read the information that has been left on it. As we have seen, however, users quickly requested the ability to access their notes from arbitrary Internet connections. They no longer used the infrared beacons as a "quick index" to the notes that were left in a particular location. Therefore, we developed an interface with a pull-down menu of locations. Users could make a selection from this list to see whether any notes had been left for them in a particular location. Early versions of this system used official room numbers to identify particular locations. This quickly proved to be unworkable because few of the people using the system even knew the official number for their own offices. We changed this to reflect the names that most people would use to distinguish the rooms: "Steve's office," "the conference room," and so on.

This approach failed to support people who were unfamiliar with the layout of the building. They preferred the official numbering system, which could be decoded to provide an indication of the floor and position of each room. The solution was to enable users to customize the names of the locations that were associated with particular infrared beacons while the system enforced a global naming system that was otherwise hidden from most users. In retrospect, it is surprising that we should not have considered this potential usability problem. In our defense, all that we can say is that such details are easily overlooked during the initial engineering of location-sensitive systems. They are also typical of the usability problems that can have a profound impact on the success or failure of such technology.

The validation of context-sensitive systems remains a research area in its own right. The mobile nature of these systems creates considerable challenges. Evaluators must track the simultaneous movements of many different users. They must also log their multiple, concurrent interactions with the system.

There are also more fundamental problems associated with identifying the sorts of tasks that might best be supported by context-sensitive systems. For example, we chose to evaluate the virtual notelet system because previous authors had identified this class of applications as an important area for the development of location aware interfaces. During the validation work, we quickly discovered that our sample population of users was relatively well served by a host of more conventional communications media, ranging from voice messaging to physical notelets to electronic mail. They had developed considerable expertise in using these systems to support their everyday tasks. Therefore, we had difficulty in persuading people to continue using the virtual notelet system. This does not imply that context-sensitive systems offer few benefits to their users. Nor does it imply that more visionary proposals, such as those embodied within the CoolTown project, will not meet with greater success. Our findings do, however, confirm the importance of obtaining direct evidence from field trials before making more elaborate claims about the potential benefits of context-sensitive computing.

THE GHOST GAME CASE STUDY

The GCS was intended to help interface designers prototype a broad range of context-sensitive applications. We have, therefore, begun to use this architecture to support the development of a multiuser system that is based on Namco's arcade game Pac-Man^{TM}. This application was deliberately chosen as a contrast to the more traditional, virtual notelet system. It was also hypothesized that a location-sensitive game might avoid some of the motivational problems that emerged during our evaluation of the previous system. The game also offered a number of additional benefits. For instance, our work with the notelet system identified a number of practical problems from our use of domestic remote controls as a tagging mechanism. These devices have numerous advantages. They are cheap, widely available, and are simple to operate. They do not require calibration. They generate a signal that can be easily detected by a vast array of existing PDAs. Domestic remote controls do, however, have a number of disadvantages.

For instance, we found that their batteries had to be recharged every 3 to 4 days during continuous use in an office environment. This imposes heavy burdens on the individuals who must maintain several dozens of these devices. It is a relatively simple task to develop a mains supply, but this violates our requirement that the GCS system should be built entirely from off-the-shelf components. We also frequently wanted to leave infrared beacons in locations where it was inconvenient or impossible to arrange for a main supply to be provided. Fortunately, the game avoided many of these problems. It was originally designed to last for an hour. We could, therefore, deploy a large number of infrared beacons without worrying about recharging the batteries or ensuring access to a main supply at all locations in the building.

The interface to the game is based on a plan of the building in which it is to be played. Rewards are then placed at strategic locations that are associated with an infrared beacon. Players win points if their PDA detects the signal of a beacon that is associated with the location in which a reward is positioned. Other beacons are then placed so that the user's PDAs can detect and report their movements within the game-playing environment. A number of the players are then nominated to play the role of ghosts. The system will remove a player from the game if a ghost's PDA transmits the identifier associated with a beacon that corresponds to the last-known location of that player. This results in situations in which a player may see a ghost approach a beacon and yet still have time to reach the range of another infrared transmitter before being removed from the game.

Preliminary evaluations have identified a host of issues that were not considered during the initial design meetings. For instance, the layout of the building has a profound impact on the gaming experience. The initial environment where we planned to host the evaluation was structured around a number of long and narrow corridors. Players had no means of escaping a ghost who was approaching down a corridor unless they were allowed to hide in rooms that were not tagged. This enabled them to hide until the ghost had moved on to another part of the building. A second design option was to allow a player to move between two different floors. This offered a further means of "escape."

This form of real-time interaction places an unusual set of demands on the Web-based infrastructure that we have described in previous sections. As noted, most mobile systems make extensive use of caching techniques to minimize communications overhead and to support disconnected operation. In contrast, the Ghost game cannot easily be played if information about the locations of other users is cached by the system. It demands prompt updates if players are to respond to the movements of their colleagues and rivals. It is, therefore, not recommended that such games should be played when other users are likely to require the radio LAN.

The rapid physical movements provoked by this game also make it advisable that the other occupants of a building are warned about the nature of any evaluation before it is conducted. Unlike many other studies into the human factors of the Web, initial trials with this application have suggested that there may be significant health and safety issues as users compete to reach the rewards that are associated with infrared beacons.

CONCLUSION AND FURTHER WORK

This chapter has described the development of a prototyping tool that is intended to help designers identify the human factors issues associated with a broad range of context-sensitive systems. The Glasgow Context Server is built from off-the-shelf hardware. This is intended to minimize the costs associated with installing the system and is also intended to simplify its maintenance. The GCS exploits domestic remote controls to provide low-powered diffuse infrared transmitters. Depending on the type of interaction that is required, these can either be used as beacons to denote particular locations in a building or as tags to help users obtain information about particular objects in their environments. This off-the-shelf approach to hardware design has also been carried over into the software components. The GCS, therefore, makes extensive use of the Web-based HTTP protocol. This approach offers numerous benefits for the development of context-sensitive interfaces:

• The Web infrastructure is ubiquitous. Both servers and browsers will run on a vast array of devices. This is important for an off-the-shelf approach to context-sensitive interaction because it should be possible to integrate users' existing PDAs into the GCS architecture. The GCS system should also avoid making strong assumptions about the hardware characteristics of the server. The Web is also ubiquitous in terms of the communications infrastructures that it can exploit. These range from conventional Ethernet networks to radio LANs and cellular architectures using the Wireless Access Protocol and similar offshoots from the HTTP.

• The Web architecture also offers more detailed technical support for the implementation of context-sensitive interfaces. The HTTP provides caching mechanisms that can be used to provide the online and offline services that can have a profound impact on the usability of any interactive application. For example, servers typically implement a HEAD operation that returns information about the status of any Web page. This can include an estimate of the time that it is safe to hold any page in a cache before it is likely to be updated. As we have seen, the designers of nomadic applications can use this data to reduce the amount of information that may have to be transferred when a user moves into and out of radio coverage. Similarly, this approach can also be used to ensure that users do not rely on cached copies of pages that store obsolete information about the location of their colleagues.

- Above all, the Web infrastructure is both simple and future proof. Therefore, it forms a strong contrast with some of the more esoteric middleware systems that have been developed to support context-sensitive applications. It is important not to underestimate the importance of these issues when interface designers attempt to expose the products of recent research to more sustained forms of usability testing.

There are, however, a number of potential problems with using the Web to support context-sensitive interaction. Some of these problems have been recognized in previous studies (Kindberg & Barton, 2001). Other limitations emerged through the development and evaluation of the virtual notelet and Ghost game case studies, which have been described in previous paragraphs:

- Minor modifications must be made to the browsers that run on end-users' PDAs. The typical Web model is one in which users' initiate the transfer of information by selecting hyperlinks or by entering URLs. In contrast, context-sensitive systems may initiate the transfer and presentation of information in response to the detection of a location beacon or an object tag. In order to make these modifications, it is typically necessary to have access to the browser's source code. This limits some of the claims made in the previous paragraphs. Although it is possible to access the Web through a vast array of devices, it can be difficult to implement the small number of changes that are necessary before a browser can be completely integrated into the GCS architecture.
- The Web-caching model does not provide adequate support for location-sensitive interaction. For instance, if an HTTP connection cannot be established, then most browsers will simply return an error code. In contrast, the GCS system must log such failed requests. It may be possible to support user tasks by repeating the request once they return to radio coverage. These logs can also be used to improve predictions about the useful content of a cache during future interaction.
- Existing Web protocols do not provide the fine-grained access control mechanisms that are necessary for nomadic interaction. It is necessary to determine whether or not a user has permission to access information or to request services from the Web-enabled objects that are an important feature of Hewlett-Packard's CoolTown proposals. Similarly, our virtual notelet study shows that even for location disclosing systems, the ubiquitous nature of Web access makes it critical that privacy is considered and maintained. It is possible to implement access control policies using the existing Web infrastructure.

For instance, .htacces files can be placed in restricted directories. This approach is error prone; users often forget to remove access restrictions or leave sensitive information unprotected. Alternatively, server administrators can alter the directory access control settings in global files, such as access.conf. This approach is inflexible. Individual users cannot easily alter the permissions that are associated with their files. It is important to remember that we are primarily interested in exploring the human factors of interaction with context-sensing systems. The fact that we must consider such details provides an ample illustration of the difficulties posed by using the Web as a principle component of the GCS. Fortunately, it is possible to hide much of this complexity by providing appropriate abstractions to interface developers. It is, however, less easy to identify abstractions that end users can exploit to dynamically alter the permissions associated with their location-sensitive information. This remains a subject of current research.

It is important to place these limitations in the wider context of recent attempts to revise the infrastructure that supports the Web. For example, improved access control and cache monitoring have a wider significance beyond the implementation of context-sensitive interfaces. Such techniques have the potential to improve the design and implementation of many different interactive, Web-based systems. It is for this reason that both issues are being directly addressed by recent initiatives to revise key components within the existing Web infrastructure. Access control is an important consideration within the document object model (DOM) that will define the logical structure of Web-based documents. The DOM model will also define the way in which a document can be accessed and manipulated.

Similarly, attempts to develop the next generation of HTTP (HTTP-NG) have described the existing mechanisms for controlling the browser cache as "not very accurate." W3C working groups have, therefore, begun to identify means of improving the performance that is implied by new generations of mobile users with sporadic connections (Spreitzer & Nielsen, 1998). These changes will have a significant impact on the usability of the web in general. By integrating HTTP and successor protocols into the GCS, we are also in a position to benefit from these developments.

We began our research with the intention that we would focus on the human-factors issues that complicate interaction with context-sensitive systems. In particular, we were anxious to validate the claims that previous authors have made about the potential benefits of these systems. To do this, we were forced to develop a test bed for conducting usability studies in this area. The GCS' integration of off-the-shelf hardware with elements of the existing Web infrastructure partially satisfies this requirement. It is not, however, a panacea. Development work continues. For instance, we have still to resolve the problem of how best to translate local, short-range transmissions into the wider addressing scheme supported by URLs. In the meantime, initial trials are beginning to reveal the problems of integrating context-sensitive services into existing working practices. We have to learn new skills as interface designers. Many of the techniques that we anticipated would provide the greatest benefits that actually provide trivial support for end-user tasks. It is arguably a strong indictment of our existing design expertise that we have had the greatest success in the development of location sensitive, multiuser games.

REFERENCES

Abowd, G. D., Atkeson, C., Hong, J., Long, S., Kooper, R., & Pinkerton, M. (1997). Cyberguide: A mobile context-aware tour guide, *ACM Wireless Networks, 3*(5), 421–433.

Baldonado, M., Cousins, S., Gwizdka, J., & Paepcke, A. (2000). Notable: At the intersection of annotations and handheld technology. In P. Thomas & H. W. Gellersen (Eds.), *Proceedings of the Second International Symposium on Handheld and Ubiquitous Computing 2000* (pp. 100–113). Berlin: Springer-Verlag.

Cheng, Y. M. (2002). *The development and evaluation of a prototyping environment for context-sensitive interaction.* Unpublished doctoral dissertation, University of Glasgow, Department of Computing Science, Glasgow, Scotland.

Cheng, Y. M., & Johnson, C. W. (2001). The reality gap: Pragmatic boundaries of context awareness. In A. Blanford, J. Vanderdonckt, & P. Gray (Eds.), *People and computers: Vol. 15. Proceedings of HCI 2001* (pp. 427–438). Berlin: Springer-Verlag.

Cheverst, K. (1999). *Development of a group service to support collaborative mobile groupware.* Unpublished doctoral dissertation, Lancaster University, Department of Computing Science, Lancaster, England.

Davies, N., Cheverst, K., Mitchell, K., & Friday, A. (1999). Caches in the air: Disseminating information in the guide system. In *Proceedings of the Second IEEE Workshop on Mobile Computing Systems and Applications* (pp. 1–19). LA: New Orleans.

DuBois, III, J. H. (1991). *A serial-driven infrared remote controller* [online]. Available: http://www.armory.com/~spcecdt/remote/ (Accessed October 15, 2001).

Ebling, M. R., John, B. E., & Satyanarayanan, M. (1998). The importance of translucence in mobile computing systems. In *Proceedings of the 15th ACM Symposium on Operating Systems Principles*: Association for Computing and Machinery.

Gardner, M., Sage, M., Gray, P., & Johnson, C. W. (2001). Data capture for clinical anaesthesia on a pen-based PDA: Is it a viable alternative to paper? In A. Blanford, J. Vanderdonckt, & P. Gray (Eds.), *People and computers: Vol. 15. Proceedings of HCI 2001* (pp. 439–456). Berlin: Springer-Verlag.

Gimson, R., Finkelstein, S. R., Maes, S., & Suryanarayana, L. (Eds.). (2001). *Device independence principles* [online, World Wide Web Consortium Working Draft, Tech. Rep.]. Available: http://www.w3.org/TR/2001/WD-di-princ-20010918 (Accessed October 15, 2001).

Harter, A., & Hopper, A. (1994). A distributed location system for the active office. *IEEE Network, 8*(1), 62–70.

Infrared Data Association. (1998). *Infrared Data Association serial infrared physical layer specification Version 1.3* [online]. Available: http://www.irda.org/ (Accessed October 15, 2001).

Ishii, H., & Ullmer, B. (1997). Tangible bits: Towards seamless interfaces between people, bits and atoms. In S. Pemberton (Ed.), *ACM SIGCHI. Proceeding of the Conference on Human Factors and Computing Systems* (pp. 234–241). Atlanta, GA: Association for Computing and Machinery.

Johnson, C. W. (1998). *Proceedings of the First Workshop on Human Computer Interaction for Mobile Devices* (GIST Tech. Rep. No. 98/1). Glasgow, Scotland: University of Glasgow. Available: http://www.dcs.gla.ac.uk/~johnson/papers/mobile/HCIMD1.html

Kindberg, T., & Barton, J. (2001). *A Web-based nomadic computing system* [online, Tech. Rep. No. HPL-2001]. Hewlett-Packard Laboratories. Available: http://www.cooltown.hp.com/dev/wpapers/nomadic/nomadic.asp (Accessed October 15, 2001).

Kindberg, T., Barton, J., Morgan, J., Becker, G., Caswell, D., Debaty, P., Gopal, G., Frid, M., Krishnan, V., Morris, H., Schettino, V., & Serra, B. (2000). *People, places, things: Web presence for the real world* (Tech. Rep. No. HPL-2000-16). Bristol, UK: Hewlett-Packard Laboratories.

Pascoe, J. (1997). The stick-e note architecture: Extending the interface beyond the user. In *ACM Proceedings of the 1997 International Conference on Intelligent User Interfaces* (pp. 261–264). Orlando, FL: Association for Computing and Machinery.

Spreitzer, M., & Nielsen, H. F. (Eds). (1998). *Short- and long-term goals for the HTTP-NG project* [online, Tech. Rep.]. Available: http://www.w3.org/TR/1998/WD-HTTP-NG-goals (Accessed October 15, 2001).

Stein, L., & Stewart, J. (2001). *The World Wide Web security FAQ* [online, Tech. Rep.]. Available: http://www.w3.org/Security/Faq/wwwsf5.html#CON-Q1 (Accessed October 16, 2001).

Want, R., Schilit, W., Adams, N., Gold, R., Petersen, K., Goldberg, D., Ellis, J., & Weiser, M. (1995). An overview of the PARCTAB ubiquitous computing experiment. *IEEE Personal Communications, 2*(6), 28–43.

Want, R., Fishkin, K. P., Gujar, A., & Harrison, B. L. (1999). Bridging physical and virtual worlds with electronic tags, *ACM SIGCHI: Proceedings of the CHI '99 Conference on Human Factors in Computing Systems: The CHI is the Limit* (pp. 370–377). Pittsburgh, PA: Association for Computing and Machinery.

14

Scent of the Web

Ed H. Chi
Xerox Palo Alto Research Center

INTRODUCTION

Scientific understandings of how people search for information are essential to the development of Web usability. The emergence of the Web as a global information environment has fueled the interests of many researchers and practitioners in how to make the Web as useful as it can be. Crucial to its success are its ubiquitous and searchable qualities. Despite its success, most of the current practices of Web design are based on empirical observations and intuition. The development of an accurate cognitive model of how users search for information on the Web is crucial to Web usability, because researchers and practitioners alike require a more complete and deeper understanding of users' decision processes than what we know today.

Information scent is the user's perception of the value and cost of accessing a piece of information. Users obtain this evaluation by perceiving, reading, and cognitively processing the various information cues that are on a Web page to decide which hyperlink to follow in order to maximize their information gain.

MOTIVATION

There exist various beliefs about Web usability—how it should be tested, examined, and improved. Many of these arguments and "rules of thumb" have been advocated without informed scientific understanding of how people actually look for information on Websites and often without any actual user experiment validation. These expert (or "guru") opinions are often based on empirical in-the-field experiences, which certainly have a place in the formation of a brand new field. However, in order to actually generalize and extrapolate from these experiences, we seek to improve our understanding of human behavior on the Web as usability specialists. The domain of understanding and modeling of human behavior is at the foundation of psychology. What can we learn from cognitive psychology to help us build better Websites?

Herb Simon, a noted psychologist and Nobel prize winner, proposed in 1957 a theory called Bounded Rationality, which states that an agent behaves in a manner that is nearly optimal with respect to its goals as its resources will allow (Simon, 1957). This proposal is interesting because it specifically states the need to understand that human agents are often suboptimal in the choices that they make and, moreover, they are limited and constrained by the amount of resources available at their disposal.

Applied to an information environment, what this theory essentially tells us is that users have limited attention available to them, and, given this constraint, they will behave in a way that maximizes the amount of knowledge they obtain in the information environment. In fact, Varian (1995) quoted Simon succinctly in a *Scientific American* article that

> What information consumes is rather obvious: it consumes the attention of its recipients. Hence, a wealth of information creates a poverty of attention, and a need to allocate that attention efficiently among the overabundance of information sources that might consume it.

The Web is a unique information environment because it is (1) large, (2) searchable, (3) ubiquitous, and (4) useful. The Web is large and growing very rapidly. Figure 14.1 depicts the growth of this information environment. One new server is introduced to the world every 2 seconds. There are currently 27.5 million Websites (Netcraft, 2001) and 413.7 million users (Computer Industry Almanac, 2000). The Web is searchable, with many search engines aiding Internet navigators. Google, for example, serves 100 million searches per day (Search Engine Watch, 2001). The Web is accessible from many different places around the world, and getting easier to access as well. Many cell phones can now be connected to information on the Web. Last but not least, the Web contains information that users want. A common phrase among Net-savvy users is "You can find the answer on the Web." The Web has fundamentally changed the augmentation of human intellect, because

Growth of WWW servers

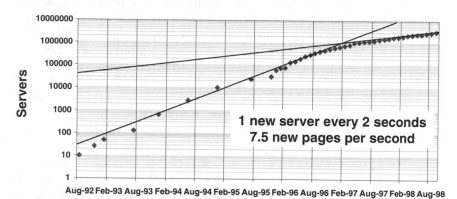

Source: World Wide Web Consortium, Mark Gray, Netcraft Server Survey

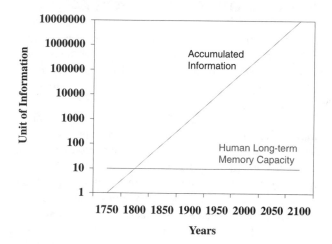

FIG. 14.1. Even though the amount of information produced per year is growing exponentially, human memory capacity in the future is not going to change (used by permission, courtesy of Stuart Card & Peter Pirolli).

this hypertext system is superior in all of those four qualities in many ways, as compared to previous information systems. The Web is larger, more searchable, more ubiquitous, and more useful than previous digital libraries.

However, even though the Web has made a wide variety of information available, this increase in the amount of accessible information actually exacerbates the problem of information access, because as humans we have limited human capacity for absorbing information. Figure 14.1 depicts this idea graphically. Human memory capacity is not likely to change in the future, but we know that the information still growing exponentially. We not only end up feeling overwhelmed (the information overload problem), but we actually need more efficient methods for allocating our attentional resource (the limited attention problem). This new perspective dramatically changes the way we should think about accessing information on the Web. It's not that we should help users digest more and more information faster, it's that we need to enable them to allocate their time on the most relevant pieces of information. This changes the goal from enabling them to access a greater quantity of information to enabling them to find higher quality information.

To increase users' efficiency and their ability to find high-quality information, we must think about models of how users search for information: What are their goals? How do they decide what to pay attention to? What is their optimization strategy? What information pops out at them? How do they skim text? How do they carry out their decision processes? How can we model that decision process?

INFORMATION FORAGING THEORY

A recent development in our research is the formulation of Information Foraging Theory (Pirolli, 1997; Pirolli & Card, 1999), which uses an ecological metaphor to understand the optimization processes that are in play in user decision making in an information environment. This theory is a novel way to think about the relationship between information goals, user contexts, and user behavior.

Information Foraging Theory developed as part of a new movement in cognitive psychology toward an *adaptationist approach* to explaining human behavior. This approach was partially developed in response to a more mechanistic explanation of human behavior, which is characterized by the Model Human Processor and GOMS model (Anderson, 1990). Mechanistic approaches construct programmed rules that simulate human performance on some task and then refine these rules by repeated engineering experiments. To some extent, this is not much better than doing straight statistical modeling of human behavior, because it only shows how behaviors happen but not why. Adaptationist approaches instead state that human behaviors arise out of evolving toward near optimal solutions in the environment. Information Foraging as an adaptationist approach states that humans optimize their strategy for the consumption of information by adapting themselves to the information environment (Pirolli & Card, 1999). The analogy is that foraging for information in information environments is the same as foraging for food in natural environments. As it turns out, the same mathematical models that describe animal foraging behavior in nature can be used to model the information foraging behavior of users in information environments.

Pirolli and Card (1999) showed that Information Foraging Theory models the user behavior in several information environments, including a browser for a set of articles (Pirolli, 1997). Our interest is to develop this model into a predictive model for Web surfing (Chi, Pirolli, Chen, & Pitkow, 2001; Chi, Pirolli, & Pitkow, 2000), especially because it has been shown to be effective in modeling other information environments.

Information Scent

A key development in Information Foraging Theory is the notion of Information Scent (Chi et al., 2000; Pirolli, 1997). Briefly defined, information scent is the user's perception of the cost and value of accessing a piece of information. What we mean by *user's perception* is that all human value judgments of whether it is worthwhile to access some piece of information are determined by the context of that user. The context of a user is determined by many different aspects, including the user's professional interest, personal hobbies, gender, news interest, personal holiday events, and monetary and time constraints. For example, a high school football player would be a lot more interested in sports medical equipment than an avid gardener.

The working context of a user is also determined greatly by the given moment. For example, a user might be an avid golfer and extremely interested in buying a new pair of golf shoes. However, if the user has a time constraint of 5 minutes before heading off to a business meeting (with a quick task of estimating of the market size for Web usability consulting), the user is not going to be interested in buying golf shoes, no matter how attractive the price. However, at a later time the user might very well be extremely interested in this new shoe sale. The value judgment of whether something is relevant at that given moment is highly dependent on the user, that is, the user's current work context and the user's overall information interest.

What we mean by the cost of accessing information is that there are different cost structures to every information environment. A poorly designed site is going to have huge cognitive cost for the user accessing it. A slow site has the users wasting large amounts of their time waiting for the server and the network to respond. A slow connection using a modem is going to have a different cost structure as compared to a fast 100-megabit Ethernet network at the office. These inherent cost structures in the environment affect the user's behavior on the Web.

What we mean by the value of information is implicit in our argument—users attach specific value to each piece of information they access on the Web. As they surf, they are continuously making these value judgments in order to efficiently navigate themselves to the most profitable patch of information. These strategies maximize their information gain.

Scientific and Engineering Understanding of Web Usability

Information scent was first applied to the Web in (Chi et al., 2000) and later expanded in Chi et al. (2001) and Heer and Chi (2001). This chapter presents the predictive model suggested by information scent. Information scent could greatly inform us about how users surf for information on the Web, and we are interested in using this notion to devise ways to make predictions of how users surf for information on the Web.

The development of this model has important implications. First, a scientific understanding of human behavior in Web surfing is fundamental to Web usability. A great deal of literature on Web usability is now available, with most based on empirical design principles and rules of thumb (Krug, 2000; Nielsen, 2000; Spool, Scanlon, Schroeder, Snyder, & DeAngelo, 1999). Certainly, experience with building Websites can inform us a great deal about what works and what doesn't. This is analogous to learning to build a log cabin through trial and error. Though possible, eventually we need to build large and complicated buildings and structures that require engineering principles to calculate the loads on the beams and optimal shapes, informing us about what's possible and what's not. The same knowledge is required in Website architecture and design. We use this model to actually make predictions about the usability of these Websites.

Second, a predictive model in Web surfing represents a new step toward Web analytics. Previous analysis of Web surfing activities are based on statistical analysis and modeling of Web traffic, typically through Web log analysis, in-lab, remote, instrumented browser log analysis. We are interested in using the model of information scent to predict surfing paths without any usage data. This moves us forward from first-order descriptive statistics to predictive statistics. A new capability, for example, would be making predictions about the usability of a Website even before it is launched live in front of real users. The development of a predictive model could greatly accelerate our ability to understand alternative designs.

APPROACH

Our approach for the development of this predictive model is first to examine the relevant computational models related to the cognition process during Web surfing, including theories of perception, visual attention, memory, decision making, problem solving, and others. Using these computational models, we develop a simulation of user behavior. We then perform an experiment measuring the behavior of users actually performing tasks using the Web. We study how much the actual user behaviors match the simulation model. Using this new knowledge of what's wrong with our original model, we modify and build better models of Web surfers. Iterating through this cycle will also enable us to think of new ways to build user interfaces that are more effective, based on our new knowledge of user behaviors in information environments.

Given this approach, our first goal is to examine users' behavior while they surf the Net. We made a number of infrastructure investments to enable remote user studies and detailed laboratory eye-tracking analysis of Web surfers (Card et al., 2001; Reeder, Pirolli, & Card, 2001). What we discovered in laboratory studies is that people typically do a quick four-step process when they perform a task on

FIG. 14.2. Users forage for information by surfing along links. Snippets provide proximal cues to access distal content. Proximal cues could be specific words that trigger action, for example.

the Web: (1) scan, (2) skim, (3) decide, and (4) act. They first scan through the Web page (often involuntarily because of visual pop-out effects) and discover anything that might grab their attention. They then skim through the potential relevant pieces of material to see if any information cues are worthwhile to pursue. That is, they decide based on the amount of information scent (Card et al., 2001). Based on relative value judgments, they decide on a particular course of action, which usually involves: (1) following a link, (2) using the back button, (3) using a search engine, (4) looking for help, or (5) quitting the task. They then actually try to carry out this decision but sometimes fail due to technological, interface, or motor control failures (e.g., back buttons disabled). Having this intimate knowledge of how people search for information enabled us to model users' behavior on the Web.

Information Scent Algorithms

Given that users typically skim and pick up information cues rapidly during Web surfing, what we are interested in is some way to automatically compute how much information cue they are receiving on each Web page. On the Web, users typically forage for information by navigating from page to page along Web links. The content of pages associated with these links is usually presented to the user by some snippets of text or graphic. Foragers use these *proximal cues* (snippets and graphics) to assess the *distal content* (page at the other end of the link).[1] Here, proximal cues refer to all the informational signals around a link that may prompt the user to choose that link. Information Scent is the imperfect, subjective perception of the value and cost of information sources obtained from proximal cues, such as Web links, or icons representing the content sources. Figure 14.2 depicts this idea.

Our assumption is that, for the purposes of many analyses, users have some information need and their surfing patterns through the site are guided by information scent. That is, we assume that users make navigational choices not randomly, but based on some rationale. The link with the highest amount of information *scent* is the link with the highest probability of being followed. Given this framing assumption, we can ask two fundamental questions, for which we have developed computational techniques:

---

[1]Furnas (1997) referred to such intermediate information as "residue."

1. Behavioral prediction: Given some user information goal and some starting point on the Web, where do users tend to go? We use Information Scent to predict the expected surfing patterns, thereby simulating Website use.

2. Need prediction: Given some user path and the associated content and linkage structure of the Web locality, what was the information need expressed by that path? For a particular pattern of surfing, we seek to infer the associated information goal based on the Information Scent that the user has followed.

The two algorithms are based on computational models of *Spreading activation* (Anderson, 1983; Anderson & Pirolli, 1984) used in the study of relevancy in human memory. Spreading activation works with a weighted graph of associations between chunks of information. Starting from an initial chunk of information (the "source"), activation spreads from one chunk to another through association links that specifies the probabilistic strength between the two chunks. Activation flows from one chunk to another much like water flowing through a system of pipes with varying widths. Spreading activation predicts that highly activated chunks of information are highly relevant to the source chunk. This process is highly similar to Bayesian inference of a probabilistic network, or a Markov-chain process. Figure 14.3 shows one depiction of this process.

The essential idea is to compute the probability of ending up at a particular node in the graph. Given that each edge specifies the probability of transition through that pair of nodes, we can compute the conditional probability of ending up anywhere in the entire graph. Applying this to the simulation of users moving through the Web, we use the nodes in this graph to represent each page. We then compute the transition probability of moving from one page to another based on the strength of the information scent. We then use spreading activation to simulate users moving through the Website with these transition probabilities.

In the following sections, we describe the behavioral prediction algorithm called Web User Flow by Information Scent (WUFIS) and the need prediction algorithm called Inferring User Need by Information Scent (IUNIS).

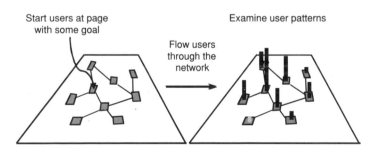

FIG. 14.3.   WUFIS and spreading activation.

Web User Flow by Information Scent

Our assumption is that, for the purposes of many analyses, users have some information goal and their surfing patterns through the site are guided by information scent. Conceptually, the simulation models an arbitrary number of agents traversing the links and content of a Website. The agents have information goals that are represented by strings of content words such as "Xerox scanning products." For each simulated agent, at each page visit, the model assesses the information scent associated with linked pages. The information scent is computed by comparing the agent's information goals against the pages' contents. This computation is a variation of the computational cognitive model for information scent developed in (Chi et al., 2001), which we describe later.

A data flow model of the algorithm is presented in Fig. 14.4. First, the entire Website is downloaded and crawled, thus extracting its content and the linkage structure. By comparing the user information need to the content and the information cues that exist on each page, we calculate the transition probability of moving from one page to another based on that information need. This transitional probability is the information scent that is collected into the scent matrix. Each

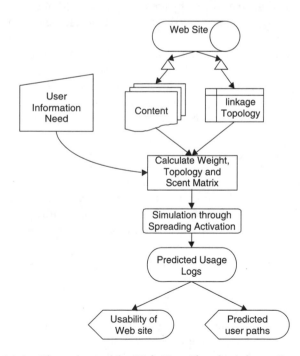

FIG. 14.4. Flow chart of the Web User Flow by Information Scent (WUFIS) algorithm.

entry (i, j) in the scent matrix specifies the probability that a user with that information need will move from page i to another page j. The spreading activation algorithm is then used to simulate users flowing through this network. The result of this simulation is then analyzed to create a set of predicted user paths and a set of metrics for the usability of the Website.

The key to the algorithm is the computation of the amount of the information scent for each link on each page. The information scent used by the simulation is computed from the amount of information cues as represented by a text snippet or icon that might stimulate a user to click on that link. There are a variety of ways to capture the amount of scent that is given off by a link. Fundamentally, the scent of a link could be calculated from analyzing: (1) the words that are in the link text itself, (2) the words in the text around that link, (3) the graphics around that link and what those graphics link to, (4) the position of the link on the page, and other characteristics. Each word is weighted by the frequency of how often this word occurs in that page and how often it occurs in the entire document collection.[2] In cases in which proximal scent could not be obtained through these computational analyses, we have also proposed to approximate the scent values by substituting the relevancy of the content of the page at the other end of the link. Presumably, the semantic of what is expressed by the link is contained in the distal document. In practice, we have found that this is always an underestimate of the true value of the proximal cues.

To summarize, at the end of this process, we will have obtained a scent matrix that describes the transitional probabilities of users clicking on a link given an information need. To simulate users flowing down these various scent conduits, we use a network flow algorithm by starting users at some page, as depicted in Fig. 14.3. For specific mathematical details of the algorithm, please see Chi et al. (2001).

Inferring User Need by Information Scent

Having discussed the behavioral prediction algorithm, we now turn our attention to the companion algorithm for user need prediction. We describe a method for inferring the information need of a user based on the user's traversal path through a hypertext collection. A user typically forages for information by making traversal decisions based on the user's task. For example, at any point in a traversal through the Website, the user has expressed interest in various pieces of information by the decision to traverse certain links. A well-traveled path may indicate a group of users who have very similar information goals and are guided by the scent of the environment. This user's traversal history is a list of documents that approximates the information need. Therefore, given a path, we would like to know the information goal expressed by that path. The question is given a traversal path through

---

[2] We use TF.IDF, a common information retrieval technique for this purpose (Schuetze & Manning, 1999, p. 543).

a hypertext collection; what can we say about the information need expressed by that traversal path?

We have developed a new algorithm called Inferring User Need by Information Scent, which uses the scent flow model in "reverse" to determine users' information goals (Chi et al., 2001; Chi et al., 2000). In order to compute the information need of a traversal path, we make the following observation. The input to the model should be a list of documents and the order in which they were visited. The output is weighted keyword vector that describes the information need. Notice that this is the direct reverse of the simulation given by the WUFIS model formulated above. The input to the WUFIS is a weighted vector of keywords that describes the information needs, and the output is a list of documents that are visited by the simulated users. Therefore, intuitively, we need to only reverse the computation for WUFIS to obtain a list of user need keywords. So it seems we need to reverse the flow of activation to obtain our result.

However, the scent matrix in the WUFIS model is computed with respect to a given information need. The scent matrix is already biased with a given information need. Therefore, we cannot simply pump spreading activation through the scent matrix, because the associated strengths are already biased with a set of keywords describing a particular information need. Therefore, to obtain our IUNIS algorithm, instead of pumping activation through the scent matrix, we need to pump activation through the link topology matrix. We can then obtain a set of keywords by examining which keywords are most important on that path. The information need expressed by the path is then described by a sorted list of weighted keywords, which can be skimmed by an analyst to estimate and understand the goals of users traversing a particular path. Figure 14.5 describes the data flow of this algorithm.

In this section, we described the two information scent algorithms at a conceptual level. For specific mathematical details of the algorithms, we refer readers to Chi et al. (2001).

Validation

We have carried out several empirical studies of the model and the algorithms. To validate WUFIS, we need to obtain usage data in which user tasks are explicitly known and coupled with user paths. This would enable us to verify whether, given some information need, WUFIS generates paths and destinations that correlate with real usage data. Unfortunately, usage data of this kind are expensive and difficult to obtain. Web server transaction logs are not sufficient because we do not know a priori what the user tasks were in those logs. All of the paths are all mixed together with different user information needs.

For a preliminary study, we wished to first validate that WUFIS generates simulated user-path destinations that are consistent with the user information needs. We chose 19 different Websites of varying size and type. The sizes ranged from

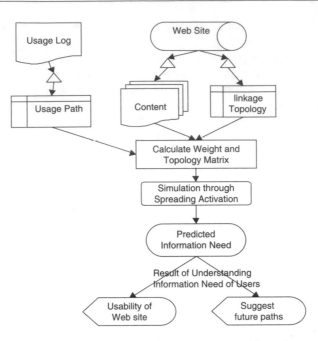

FIG. 14.5. Flow chart for the Inferring User Need by Information Scent (IUNIS) algorithm.

27 to 12,000 pages, while site types ranged from informational to e-commerce and large corporate sites. We also varied the information need query vector from very general (looking for product information) to very specific (migraine headaches). The ending position of simulated paths were generated, and the top 10 URL destinations were extracted, as they represented the most relevant user endpoints to the information need according to the algorithm. A human judge rated each [query, site, and URL]-tuple from a scale of 0 to 10, with 10 meaning that the URL is "most relevant" to the query and 0 meaning "not relevant at all." These ratings were done blindly so that the judge did not know which algorithm produced which URL. After all 190 rating were collected, the results were tabulated and averaged to obtain the ratings for the algorithm for each site. Table 14.1 shows the result of this preliminary experiment. The destinations generated by WUFIS were mostly accurate except in four cases in which the rating fell below 7. Given that the number of possible paths in a Website is huge, we were pleased that WUFIS generated path destinations that are rated fairly highly.

We also wished to show that IUNIS produces a set of keywords that are indicative of the core semantic concepts in users' information need as expressed by their surfing path. To do this, we performed an evaluation of the IUNIS-generated keyword summaries with respect to their ability to communicate the content of user paths. We randomly chose 10 paths from a large Xerox.com Web log and

TABLE 14.1

Analysis of WUFIS Algorithm Results on 19 Different Web Sites of Varying Sizes and Types

Web Site	Category	#Docs	#Words	Information Need	Rating
ArtShow.com	Museum/Gallery	1,063	13,655	Landscapes, Painting, Paintings, Landscape	7.3
Achoo.com	Informational	770	11,568	Migraines, Headache, Headaches, Treatment, Drugs	4.5
Healthfinder.gov	Informational	4,684	68,742	Migraine, Headache, Headaches, Treatment, Drugs	5.1
Healthgate.com	Informational	2,921	25,743	Migraine, Headache, Headaches, Treatment, Drugs	7.3
medweb.emory.edu/ MedWeb	Informational	11,284	15,488	Migraine, Headache, Headaches, Treatment, Drugs	8.2
Inxight.com	Small Company	207	6,827	Product, Products, Reproduct, Productivity	7.4
Aircourier.com	Small Company	27	933	September, Mexico, Cheap, Airfare	8.1
Loudcloud.com	Small Company	103	5,624	Product, Products, Productivity, Reproduct	8.3
Snapple.com	Small Company	64	7,957	Twisted, Cap, Promotion, Promotions	8.7
Target.com	E-commerce	2,836	13,001	Clothes, Clothing, Accessory, Accessories, Apparel	9
Kmart.com	E-commerce	860	9,152	Clothes, Clothing, Accessory, Accessories, Apparel, Shirt, Shorts	9
PaloAltoDailyNews. com	Online news	412	14,985	Housing, Homes, Shortage, Homelessness, Homeless	8.4
Newsweek.com	Online news	1,969	61,414	Democratic, Issues, Gore, Lieberman	9
Trailplace.com	Message Board	232	72,786	Trail, Trails, Trailheads	7
VolunteerMatch.org	Nonprofit/Search	12,054	42,697	Children, Youth, Palo, Alto, Ca	4.4
Julliard.edu	Education	114	12,725	Admission, Admission, Applications, Applications, Procedure	8.5
Cs.umn.edu	Education	4,928	54,070	Collaborative Filtering	9.4
Canon.com	Large Corporation	1,254 (depth 8)	14,652	Printer, Printers	6.4
Xerox.com	Large Corporation	2,218	21,819	Product, Products	7.3
Average					7.54

constructed booklets containing the documents on these paths. A single rating sheet containing the keyword summaries from IUNIS for each of the corresponding user paths was given to each of the 8 participants. On a 5-point Likert scale, each participant rated the relevancy of each of the 10 summaries to each of the 10 user paths for a total of 800 ratings (8 participants × 10 summaries × 10 user paths). The keyword summaries generated by IUNIS for a particular path (matching summaries) were rated more relevant than the summaries generated for other paths (nonmatching summaries). On the scale "1 = not relevant" to "5 = highly relevant," the matching summary mean was 4.58 ($Mdn = 5$), and the nonmatching summary mean was 1.97 ($Mdn = 1$). This difference was highly significant: $F(1,781) = 283.08$, $MSE = 1.73$, $p < .001$. This indicates that the path keyword summaries generated by the IUNIS algorithm were judged to be very good representations of the path content.

The participants were also asked to choose one summary as the best match to each of the 10 user paths. An even stronger test was provided by the participants' selection of "the best summary" for each of the paths. On average, participants chose the IUNIS summary as the best match 5.6 times out of 10 ($SD = 1.3$; chance selection $= 1$ time out of 10). A measure of the degree of match was provided by Cohen's kappa statistic, which ranged like a correlation coefficient (in this case, kappa $= 0$, indicates random association, and kappa $= 1$, equals perfect association). Cohen's kappa $= 0.51$ for the degree of match between participants selections of "best summary" and the IUNIS summary. This indicated a good match between participants' selection and IUNIS.

These two evaluations gave us confidence in the general model proposed by information scent. WUFIS generated user traversal destinations that are consistent with the user information need, whereas IUNIS generated keyword summaries that are in agreement with the expressed information needs of the user paths. In the future, we will be performing larger scale validations to gain further trust in the information scent model.

Application

Each day, Web users generate over a billion clicks through the myriad of accessible Websites. Naturally, Website designers and content providers seek to understand the information needs and activities of their users and to understand the impact of their designs. Given the magnitude of user interaction data, there exists a need for more efficient and automated methods to analyze the goals and behaviors of Website visitors and analyze and predict Website usability. We have built several Web analysis applications based on the information scent model to solve both of these problems. In this section, we describe two of these applications:

1. One application is to cluster user information needs that are discovered by examining user paths from Web server transaction logs (Heer & Chi, 2001). The advantage here is that IUNIS can extract user information needs from the logs more accurately than other methods.
2. The other application is to automatically predict user paths using the simulation algorithm, thereby performing a usability analysis of the Website. By automating a Website usability analysis without actual experiment participants, we can dramatically reduce the cost of performing usability tests.

CLUSTERING OF USER
INFORMATION NEEDS

Web usage mining has been a hot research topic in the last several years due to its implications for businesses on the Web. By now, owners of Websites realize that the usability of their site could greatly determine the success of their businesses (Nielsen, 2001). Understanding user behaviors on Websites enables site owners to make sites more usable, ultimately helping users to achieve their goals more quickly. For example, this information could help Web masters prioritize the navigational paths of the static content pages to optimize for more common tasks (Yan, Jacobsen, Garcia-Molina, & Dayal, 1996). Alternatively, they could use this information to personalize content for their users (Barrett, Maglio, & Kellem, 1997). Marketers could present more relevant and targeted advertisements or sale promotions. News sites would like to produce and present materials that are highly relevant to their visitors. Server performance experts could use this information to enhance server performance by determining which features are used the most often. The aim is to make the site "stickier" so that users stay longer because of enhanced experiences. Several workshops have been organized to discuss the issues surrounding the various stages of Web usage mining (SIAM01, 2001; WebKDD01, 2001).

Using Website transaction logs, there have been many different approaches to categorize and cluster user interest profiles into groups for these purposes (Cooley, 2000; Heer & Chi, 2001). These techniques build user profiles by combining the navigation paths representing the visit sessions with other data features, such as page view time, hyperlink structure, and page content.

We have combined these data features with IUNIS keyword summaries to cluster user sessions (Heer & Chi, 2001). Multi-Modal Clustering (MMC) is a technique that utilizes multiple information data features (modalities) to produce clusters. We first collect the Content, Usage, and Topology (CUT) data of the Website to be analyzed. Next, we create a representation (or profile) of user interests based on the pages that lie on each user's surfing session. We then define a similarity metric for the user profile vectors. Finally, we cluster the user profile vectors to obtain categorizations of the profiles.

A major problem in applying user session clustering techniques is that there has been no systematic evaluation of the approaches taken. We do not know how well each of these data features contributes to the clustering process in real-world situations, because each clustering paper describes case studies using different data sets. What's worse is that there is no way to know how accurate these findings are, because without knowing a priori what users' tasks and information needs are, we are incapable of determining whether the technique really correctly clustered these user sessions into the right groupings. To do an effective evaluation, we need user sessions in which we know the associated information goals, which then enable us to evaluate whether the clustering algorithms correctly classified these user sessions into good groups.

We conducted a user study and a systematic evaluation of clustering techniques. We first asked about 20 users to surf a given site with specific tasks. This allowed us to know a priori what the user information needs of the users were. We then used this a priori knowledge to evaluate the different clustering schemes.

What we discovered was that, by counting the number of correct classifications, combining IUNIS keyword summaries with other data modalities enabled us to get accuracies of up to 99%, which is a considerable improvement over the 74% accuracy of the traditional naive scheme. Our experiment shows that clustering user sessions should be done carefully so that designers do not use wrong conclusions to make optimization decisions. More importantly, we were able to obtain extremely high accuracy by paying attention to the data modalities used in the clustering process. This is encouraging news for people trying to understand site usage.

We conducted a case study on the current www.xerox.com server access logs from July 24, 2001. Figure 14.6 depicts the results of the case study. The most striking result was that a large segment of user sessions (41.4%) center around the splash page. Viewing the actual clustered paths revealed that these sessions consisted of many paths jumping between the splash page and other linked pages. This could indicate that a large segment of users may come to the site without well-defined information needs and/or that the site may suffer possible usability problems that prevent users from successfully moving deeper into the site.

Other substantial groupings include Xerox product catalog browsing (15%), driver downloads (13.9%), technical support (11.5%), and company news and information (8.2%). One unexpected result was that there was a strong, concentrated group of German users that necessitated a unique cluster (1.7% or 7750 sessions).

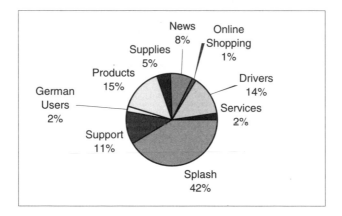

FIG. 14.6. Clustering user sessions (Data from Xerox.com, 7/24/2001).

Xerox's sales and marketing might also be interested to know the number of online shopping- and purchase-related sessions (1.3%, or 5,716 sessions) in comparison to the number of product catalog viewers. This case study illustrates that these user session clustering methods are indeed applicable to real-world situations and are scalable to large, heavily used Websites.

BLOODHOUND: AUTOMATED SITE USABILITY ANALYSIS

Designers and researchers of Web interactions have been seeking ways to quantify the quality of user experience. Naturally, certain tasks are better supported by these Websites than others, and certain information needs are satisfied more easily than other needs. Beyond log analysis and in-the-lab usability testing for this purpose, we are seeking ways to predict and measure usability of a Website by using the WUFIS simulations.

Specifically, we sought to answer questions concerning the entire Website, specific pages, and the users. What are the predicted surfing traffic routes (e.g., branching patterns or pass-through points)? How does the site measure on ease of access and cost? Given a page in the Website, what other pages are possible destination points based on a specific information need? Do actual usage data match these predictions and why? What is the cost (e.g., in terms of download time) of surfing for these visitors?

We have built an automated analysis system for the predictive modeling of Web usability of Websites. The idea behind the Bloodhound project is to create a service that would automatically infer the usability of a Website via simulated users that surf for specific information goals. The application scenario is that the service user would specify the Website and the information goal to be simulated in the analysis. The information goals are specified as a set of keywords of user information needs. Then the Bloodhound service would return usability metrics that specify how easy it is to accomplish the information goal that was given on that site. In essence, Bloodhound takes as input a Website address and a set of user tasks and returns a usability report. Figure 14.7 shows the concept behind Bloodhound.

Bloodhound has been in development for about a year and represents a significant departure from traditional usability testing methods. We have shown that the system can be used to: (1) discover pages that act as branching points or conduits, (2) discover well-traveled paths or significant highways in the Website, (3) identify possible information needs based on surfing paths, and (4) identify potential destinations given an information need (Chi et al., 2000).

Although WebCriteria has an agent surfing analysis tool that measures accessibility based on hyperlink structure of the site and the amount of content, an analysis of the actual content is not performed. As the only related system to ours, by comparison it has no representation or understanding of users' informational

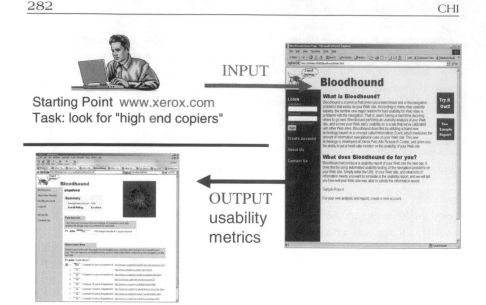

FIG. 14.7. Bloodhound automatically measures the usability of a Website.

goals (Pirolli, 2000). By contrast, the technology behind Bloodhound has a representation of users' information need and surfs according the information cues that exists on the Web pages of the site. We envision various uses of this technology:

• Current approaches to Web site analysis are aimed at the Web masters who are interested in exploring questions about the current design of a Website and the current set of users. However, Web masters must also be interested in predicting the usability of alternative designs of their Websites. They also seek to answer these same questions for new kinds of (hypothetical) users who have slightly different interests from the current users.

• Since this technology requires no human participant evaluations in the loop, the analysis can be done cheaply and repeatedly to track the development and evolution of a Website. This tracking can be done for both sites that are currently being developed or for a site that has been launched but is constantly evolving.

• Bloodhound can be used to find Websites with poor usability that need improvement (e.g., a lead-generation tool for Web usability consultants).

CONCLUSION

Web usability, as a field, has grown tremendously in the last few years. There are now a large number of practitioners out there who have experiences of what seems to work and what seems to fail. Many "usability guru" opinions and rules of thumb are passed around in the field as important design rules. We believe that what is needed instead is a scientific systematic understanding of how users surf for information.

In this chapter, we described an approach based on the concept of information scent to simulate the navigation behaviors of users on a Website with some given information need. The approach is based on the observation that Web users utilize various information cues near a hyperlink to decide whether the link is worthwhile to follow. Information scent is the measurement of the amount of information cues that are given off by a hyperlink that is likely to be relevant to the user's information need. The two algorithms described in this chapter can predict the behavior of users on a Website based on the given information need and infer the information goals of the users as expressed by their user paths.

There is a huge predicted utility to being able to simulate Website usage and infer user information goals. We have used this research to understand the usability of a Website. We demonstrated two applications that utilize this approach.

One application uses the simulation algorithm to predict the behavior of users on a Website and produces a usability report based on the prediction. Previous approaches in Web usability analysis typically involve analysis of existing user data on a current version of the site. With a scientific understanding of a user's cognitive decision process during navigation, we are able to predict the usability of a Website before it is even launched in front of real users. By developing models that are verified, we are also able to do usability testing of Websites without having to incur the high cost of employing human participants.

Another application infers the information needs of users by examining the scent tracks they followed as recorded by Web server logs. As verified in a user study, this application is potentially 99% accurate in recovering the information need of the user sessions. By helping Website analysts understand the information goals of users who visit the site, we help them understand how well the site is supporting the users in achieving their goals. The continual development of cognitive models of Web surfing will enable researchers to better understand the usage of the Web, designers to better design their Websites, and end users to seek information more efficiently.

ACKNOWLEDGMENTS

The work described here are the results of collaboration with Peter Pirolli, Jeff Heer, Jim Pitkow, Kim Chen, Stu Card, the User Research Interface (UIR) research group, and the Bloodhound Project Team in the ASD/Y-Axis development group at Palo Alto Research Center (PARC). I wish to thank Pam Schraedley for comments and suggestions on an early draft of this chapter. This research was supported in part by an Office of Naval Research grant no. N00014-96-C-0097 to Peter Pirolli and Stuart Card.

REFERENCES

Anderson, J. R. (1983). A spreading activation theory of memory. *Journal of Verbal Learning and Verbal Behavior, 22,* 261–295.

Anderson, J. R. (1990). *The adaptive character of thought.* Hillsdale, NJ: Lawrence Erlbaum Associates.

Anderson, J. R., & Pirolli, P. L. (1984). Spread of activation. *Journal of Experimental Psychology: Learning, Memory, and Cognition, 10,* 791–798.

Barrett, R., Maglio, P. P., & Kellem, D. C. (1997). How to personalize the Web. In *Proceedings of the ACM Conference on Human Factors in Computing Systems, CHI '97* (pp. 75–82). Atlanta, GA: Association for Computing Machinery.

Card, S. K., Pirolli, P., Van Der Wege, M., Morrison, J., Reeder, R. W., Schraedley, P., & Boshart, J. (2001). Information scent as a driver of Web behavior graphs: Results of a protocol analysis method for Web usability. In *Proceedings of the ACM Conference on Human Factors in Computing Systems, CHI 2001* (pp. 498–505). Seattle, WA: Association for Computing Machinery.

Chi, E. H., Pirolli, P., Chen, K., & Pitkow, J. (2001). Using information scent to model user information needs and actions on the Web. In *Proceedings of the ACM Conference on Human Factors in Computing Systems, CHI 2001* (pp. 490–497). Seattle, WA: Association for Computing Machinery.

Chi, E. H., Pirolli, P., & Pitkow, J. (2000). The scent of a site: A system for analyzing and predicting information scent, usage, and usability of a Web site. In *Proceedings of the ACM Conference on Human Factors in Computing Systems, CHI 2000* (pp. 161–168). The Hague, the Netherlands: Association for Computing Machinery.

Computer Industry Almanac. (2001). Available: http://www.c-i-a.com/

Cooley, R. (2000). *Web usage mining: Discovery and application of interesting patterns from Web data.* Unpublished doctoral dissertation, University of Minnesota, Minneapolis.

Furnas, G. W. (1997). Effective view navigation. In *Proceedings of the Human Factors in Computing Systems, CHI '97* (pp. 367–374). Atlanta, GA: Association for Computing Machinery.

Heer, J., & Chi, E. H. (2001). Identification of Web user traffic composition using multi-modal clustering and information scent. In Jaideep Srivastava & Joydeep Ghosh (Eds.), *Proceedings of the Workshop on Web Mining, SIAM Conference on Data Mining* (pp. 51–58). Chicago: SIAM Press.

Krug, S. (2000). *Don't make me think: A common sense approach to Web usability.* Indianapolis, IN: New Riders.

Netcraft. (2001). Available: www.Netcraft.com

Nielsen, J. (2000). *Designing Web usability.* Indianapolis, IN: New Riders.

Nielsen, J. (2001, August 19). *Jakob Nielsen's Alertbox* [Online]. Available: http://www.useit.com/alertbox/20010819.html

Pirolli, P. (1997). Computational models of information scent—following in a very large browsable text collection. In *Proceedings of the ACM Conference on Human Factors in Computing Systems, CHI '97* (pp. 3–10). Atlanta, GA: Association for Computing Machinery.

Pirolli, P., & Card, S. K. (1999). Information foraging. *Psychological Review, 106,* 643–675.

Pirolli, P. (2000, March). A Web site user model should at least model something about users [Online]. *Internetworking, 3.1.* Available: http://www.internettg.org/newsletter/mar00/critique_max.html

Reeder, R. W., Pirolli, P., & Card, S. K. (2001). WebEyeMapper and WebLogger: Tools for analyzing eye tracking data collected in Web-use studies. In *Proceedings of the ACM Conference on Human Factors in Computing Systems, CHI 2001* (pp. 498–505). Seattle, WA: Association for Computing Machinery.

Schuetze, H., & Manning, C. (1999). *Foundations of statistical natural language processing.* Cambridge, MA: MIT Press.

Search Engine Watch. (2001). Available: www.searchengineatch.com

[SIAM01] (2001). Jaideep Srivastava & Joydeep Ghosh (Eds.), *Proceedings of the Workshop on Web Mining, SIAM Conference on Data Mining.* Chicago.

Simon, H. A. (1982). *Models of bounded rationality.* Cambridge, MA: MIT Press.

Simon, H. A. (1957). *Models of man.* New York: Wiley.

Spool, J. M., Scanlon, T., Schroeder, W., Snyder, C., & DeAngelo, T. (1999). *Web site usability.* San Francisco: Morgan Kaufmann.

Varian, H. (1995, September). The information economy. *Scientific American,* 200–201.

[WEBKDD01]. Ronny Kohari, Brij Masand, Myra Spiliopoulou, & Jaideep Srivastava (Eds.), *Proceedings of the SIGKDD Workshop on Web Data Mining* (WEBKDD01). San Francisco: Association for Computing Machinery.

Yan, T. W., Jacobsen, M., Garcia-Molina, H., & Dayal, U. (1996). From user access patterns to dynamic hypertext linking. *Computer Networks, 28,* 1007–1014.

15

Live Help Systems

Johan Aberg
Nahid Shahmehri
Linköpings Universitet, Sweden

An important challenge for universal usability is to bridge the gap between what users know and what they need to know to successfully interact with a computer system (Shneiderman, 2000). Different users have different needs and skills, especially for Web information systems (WISs; Isakowitz, Bieber, & Vitali, 1998). WISs often serve an international user community aiding people with various backgrounds. Thus, there is a strong need to provide support for the whole range of users. Ben Shneiderman identified online help and customer service as two important approaches in dealing with this challenge and called for evaluations and design guidelines (Shneiderman, 2000). Earlier, Carroll and McKendree (1987) argued that empirical studies on advice-giving systems should be done. Online help means online support for the user tasks that the WIS is intended for. We see the term *customer service* as the specialization of online help to the task of online shopping. In this chapter, we consider online help and use the terms *user support*, *assistance*, and *help* interchangeably.

It has been shown that user support has an influence on WISs such as electronic commerce sites. For example, the study by Vijaysarathy and Jones (2000) suggests that customer service for the preorder part of the shopping process has a positive influence on user attitudes toward Internet catalogue shopping. Also, Fogg et al.

(2001) showed that customer service is of great importance for the credibility of a Website.

Still, the current state of practice in user support for WISs is limited and in need of improvements. In a study of consumer reactions to electronic commerce by Jarvenpaa and Todd (1997), it is shown that user support is limited in electronic commerce systems. In the study by Spiller and Lohse (1998), several aspects of customer service were analyzed. The result was that for the stores under study, customer service was limited. As further motivation for studying user support for WISs, we consider the Home Net field trial, conducted by Kraut and colleagues. The results from this study showed that much user support is needed for Internet usage in general (Kraut, Scherlis, Mukhopadhyay, Manning, & Kiesler, 1996).

Given this situation, with more-or-less nonexistent user support on the Web, it is important to consider how the situation could be improved. A promising kind of user support system for Websites is the so-called *live help system*. The main characteristic of a live help system is that it involves human assistants. Similar in spirit to a traditional call center, a live help system supports users of a Website by letting them get in contact with human assistants and have consultation dialogues online. As we have discussed earlier (Aberg & Shahmehri, 2000), human assistants can potentially provide very flexible user support because of their human intelligence and background knowledge. They may also give the support a human touch.

LIVE HELP SYSTEM

In our work on live help systems we have focused on a collaboration between human assistants and computer-based support. Following an early analysis of the advantages with such a collaboration, we conducted a proof of concept study, reported in Aberg and Shahmehri (2000). The results were encouraging, and we have now carried out a larger follow-up study, on which this chapter is based.

An overview of our live help model is presented in Fig. 15.1. The support router is responsible for deciding whether the user needs computer-based support or

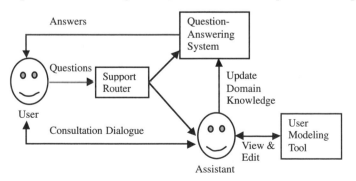

FIG. 15.1. Overview of the live help model.

support by a human assistant. The computer-based support is a question-answering system. If the support router connects the user to a human assistant, they can have a real-time consultation dialogue. A user modeling tool for supporting the assistant is also part of the model.

In our implementation of the model, the support router always routes the user through the question-answering system before connecting to a human assistant. The question-answering system is implemented using an information retrieval approach with frequently asked questions (FAQs; Aberg & Shahmehri, 2001a). The knowledge base of FAQ items is continuously updated by the assistants. Following a consultation session in which a new type of question was asked, the assistant is supposed to formulate a new FAQ item and add it to the knowledge base. This way, the automatic question-answering system can gradually improve its performance and take over more and more work from the human assistants.

Regarding the user modeling approach, information about a user is stored in a predefined attribute hierarchy, in an overlay style. A user's model is displayed for an assistant as soon as a consultation dialogue begins. The assistant can then make use of the information in the model to tailor the consultation to the individual user. No automatic inference is made on the data in the user model, although the assistant is of course free to make inferences as a part of his or her interpretation of the user data. The assistant can also update the model by filling in attribute values based on what is learned from the consultation dialogue with the user. Further, some basic demographic information (age, gender, and country) is automatically inserted in the user model via questions in a registration phase for the live help system (not shown in Fig. 15.1). More information about the user modeling system is presented in Aberg, Shahmehri, and Maciuszek (2001).

The user's support process is initiated when the user asks a question in natural language. The question is fed to the automatic question-answering system. FAQ items that closely match are returned as potential answers to the question. If the user indicates that the returned answer is not satisfactory, the support router will connect the user to a human assistant with expertise matching the topic of the question. If all the appropriate assistants are currently busy, the user can choose to wait in a queue. Once an assistant is available, the user is connected to that assistant and can proceed with a consultation dialogue via text chat.

The interface for users is illustrated in Fig. 15.2. The lowest text field with a white background is the input field. Here the user can type in a question and later type in messages to the assistant. The middle text field displays the conversation history. The uppermost text field displays system information. This includes information about the activities of the assistant, such as whether he or she is currently typing a message or searching for information to help the user. At the very top of the window is information about the number of users and assistants that are currently logged in to the live help system. There is a distinction between the assistants that are logged in and the assistants that are available. An assistant is not available when he or she is chatting with a user. The main interface for assistants

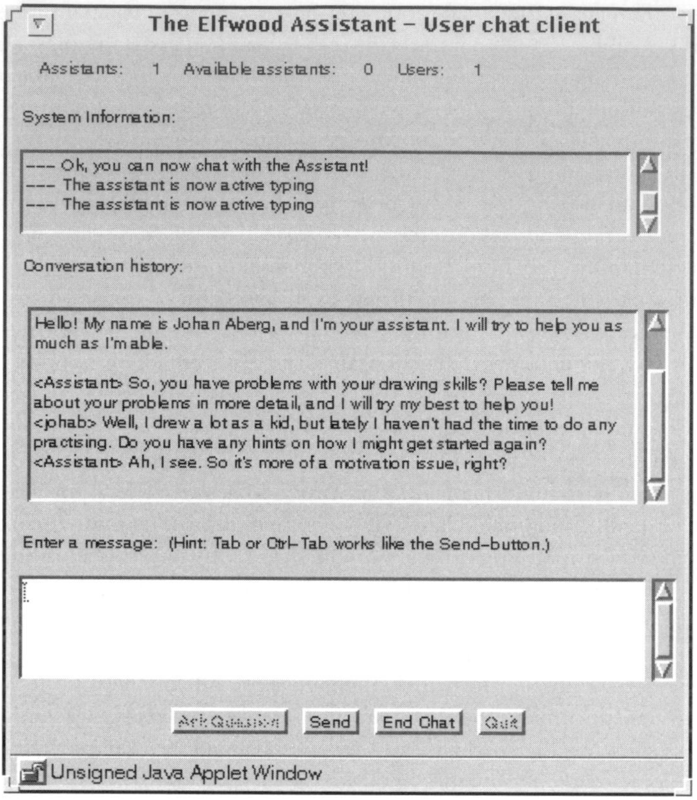

FIG. 15.2. Screen shot from the user interface.

is basically the same, but with additional buttons for opening windows with the
user model editor and the FAQ editor, and a few other details.

FIELD STUDY

Issues

Our focus in this study is on the following questions: How does the live help system
affect the attitude of users toward a WIS? Can the system provide efficient user
support? What is the work situation for human assistants like?

    When it comes to user attitude, we consider the issues of trust, fun factor, and
atmosphere. Trust is an aspect of attitude that has recently received much attention
in the literature. Evidence on the importance of trust for electronic commerce is
presented in Fogg et al. (2001), Lee, Kim, and Moon (2000), and Ratnasingham
(1998), and is discussed in the context of GroupWare in Rocco (1998). Although

the effect of trust on customer loyalty has been studied previously, it is important to also study how trust can actually be improved. Thus, it is interesting to study how human assistants affect users' trust for a WIS.

The fun factor and atmosphere of a WIS are highlighted in Jarvenpaa and Todd (1997). The study shows that users of online shops miss the fun and store atmosphere of shopping in traditional stores. Thus, it would be of interest to improve these attitude factors.

Regarding efficiency, we consider issues such as flexibility, quality of support, feasibility of textual chat, and the feasibility of having users waiting in a queue for human assistance. Knowledge gained about these issues is important for the successful deployment of this kind of support system. It is also necessary to consider the work situation of assistants to know how to support them in the best way and what is required of an assistant.

Environment

The user support system was attached to an existing WIS for a period of 3 weeks. The WIS, called Elfwood, is a nonprofit WIS in the art and literature domain, where amateur artists and writers can exhibit their material. At the time of this writing, around 6,700 artists and 1,300 writers exhibit their work in the fantasy and science fiction genre. The artists and writers are members of the site and have access to an extranet for updating their own exhibition areas. A part of the site is devoted to teaching art and includes a large number of feature articles on different topics.

Elfwood has around 14,500 daily visitor sessions (many of these are by nonmembers) in which each session averages approximately 35 minutes. About 60% of the sessions are by users from the United States. The remaining users are mainly from Europe and Canada.

There were three main user tasks that we intended to support: (1) member activities, such as uploading new art and literature, and the management of each member's exhibition area; (2) searching for interesting art and literature at Elfwood; and (3) learning how to create fantasy and science fiction–related art and literature.

We chose Elfwood as the environment for our study for two reasons. First, we wanted a site with a reasonable number of users and user traffic, and with a user community that would allow the recruitment of suitable assistants. Second, we wanted to test our system in a low-risk environment where unexpected system problems would not have large financial consequences. Based on these requirements, we considered Elfwood to be a good environment for the field study.

Participants

Voluntary assistants participated in the study (without payment) from their home or work environment. They were recruited some months before the field study began, and they were all Elfwood members. A recruiting message was posted on the site with a description of the role and requirements of an assistant. In the end, 35 people with proper expertise participated actively as assistants throughout the period of the field study, and 30 of these helped users (the other 5 were only online at times when not so many users needed support). A schedule with the online times of the assistants for the entire field study period was established and published at the site.

Before the field study started, we advertised the existence of the support system at Elfwood in a message column on the main page, and by sending an e-mail to all members. During the field study, 636 users registered with the support system, and 129 of these users worked through the system to have help conversations with assistants. Once the study started, we had an eye-catching icon on the main page that worked as a link to the help system page. To further facilitate the use of the online help system, we also had icons that linked to the help system from every page at the site. Unfortunately, our strategy for attracting users' attention did not work out as well as we hoped. After the field study, we heard comments from many Elfwood regulars that they had completely missed the existence of the live help system during the field study period.

Data Collection

We chose to use questionnaires as our evaluation tool. We saw questionnaires as appropriate for our focus on the subjective opinions of assistants and users. The use of alternative evaluation tools such as interviews and think-aloud evaluations (Karat, 1997) were considered to be difficult, because of the wide geographical distribution of the users and assistants. We used three questionnaires, and in the design of these we considered the guidelines in Bouchard (1976).

User Questionnaire 1

The first user questionnaire considered users' subjective opinions on aspects of attitude and efficiency. In order to get a reasonably high response rate, we limited the number of statements and questions as much as possible. Regarding attitude, we considered trust, the fun factor, and atmosphere. All three aspects of attitude are mentioned as relevant in related literature (as discussed in the introduction). We also considered how users would react if there were no assistants available when they needed help. Regarding efficiency, we considered ease of use, quality of support, users' view of queuing for human support, the feasibility of textual chat as a communication means, users' usage purpose, and general system flexibility.

The questionnaire contained a number of statements and questions related to the mentioned aspects of efficiency and attitude. The respondents were asked to express their agreement or disagreement with each statement by giving a rating on a scale of 1 to 10. The questions had alternative answers that users were told to choose from. The respondents were asked to explain their answers to all statements and questions. They were also asked to consider the concept of the support system in general and not focus on the actual implementation. The questionnaire was sent out to the 129 users who had had help conversations with assistants. It was sent via e-mail just after the field study was finished. We received 53 questionnaires that were properly filled out (a response rate of 41%).

For detailed demographics of the respondents for all three questionnaires, please refer to Aberg and Shahmehri (2001b). The respondents had 1.55 help conversations on average.

User Questionnaire 2

During the course of the field study, we noticed that many users had registered with the support system but never had a help conversation with an assistant. In order to investigate the reasons for this behavior, we designed a second questionnaire directed toward these users. Several possible reasons were provided, together with the option of creating alternative reasons. The questionnaire was sent out to all 507 users who had registered but not participated in a help conversation. It was sent out directly after the field study ended. We received 175 answers (a response rate of 35%).

Assistant Questionnaire

The questionnaire that was sent out to the assistants considered the work situation for assistants. Several questions were used to address this issue. We chose to use open-ended questions for two reasons. First, we believed that the assistants would be likely to spend the extra time needed for these questions compared to, for example, rating statements. Second, we were looking for as much detailed information as possible, making open-ended questions a good alternative (Karat, 1997).

Directly after the field study period had ended, the questionnaire was sent out to the 30 assistants who had been involved in helping users via e-mail. We received 21 properly filled-out questionnaires (a response rate of 70%). On average, a responding assistant participated in 6.7 help conversations.

FINDINGS

User Questionnaire 1

The results for the statements in the questionnaire are presented in Table 15.1. Apart from the statements, the questionnaire also contained two questions. The results from these questions are shown in Table 15.2 and Table 15.3. Notice that in Table 15.2 some users saw more than one alternative as appropriate. This is

TABLE 15.1

Results From Statements in User Questionnaire 1

Questionnaire Statements Regarding Efficiency	M	SD
1. To get help from this kind of assistant system is (1 = complex, 10 = straightforward)	7.42	1.94
2. The flexibility of this kind of assistant system is (1 = low, 10 = high)	7.40	1.68
3. Waiting in a queue for a human assistant is (1 = useless, 10 = worthwhile)	7.42	2.28
4. The time I would be willing to wait in a queue for an assistant is (1 = short, e.g., max. 1 minute; 10 = long, e.g., max. 20 minutes)	5.51	2.33
5. The quality of support given by assistants in general is (1 = bad, 10 = good)	7.91	1.83
6. Having a dialogue with an assistant via a textual chat is (1 = cumbersome, 10 = straightforward)	8.32	1.85
Questionnaire Statements Regarding Attitude		
7. Getting support from this kind of assistant system is (1 = boring, 10 = fun)	8.04	1.90
8. My trust in the provided support is (1 = low, 10 = high)	8.58	1.56
9. The effect that human assistants have on my trust for a Web site is (1 = negative, 10 = positive)	8.83	1.45
10. The effect that human assistants have on the atmosphere of a Web site is (1 = negative, 10 = positive)	9.08	1.37

TABLE 15.2

Results on Reactions to a System With No Assistants Logged In

Reactions to a System With No Assistants Logged In	
A. *Try again later*	22 (42%)
B. *Never come back*	3 (6%)
C. *Make use of the FAQs*	21 (40%)
D. *Try to solve the problem on one's own*	15 (28%)

TABLE 15.3

Results on Different Usage Types

Different Kinds of Usage	
A. *To save time and energy*	17 (32%)
B. *To get help with something difficult*	7 (13%)
C. *Both of the alternatives (A and B)*	27 (51%)

why the percentages sum up to more than 100%. Also, concerning the results in Table 15.3, two respondents did not find any of the alternatives fitting, which is why the percentages do not quite sum up to 100%.

We summarize the results from user questionnaire 1 in light of the explanations that we received using content analysis. The following method was used in analyzing the respondents' explanations. First, to get a good overview, we created a file for each statement where all explanations were stored verbatim. Second, for each statement, we made a summary of all issues that were mentioned in the explanations and noted the number of times that they were mentioned. Third, issues that were closely related were organized into categories with corresponding descriptive terms, keeping track of the number of times that each category was mentioned. Note that this step sometimes required interpretation. Fourth, for the sake of brevity, only the most informative categories were selected for the presentation. Still, we attempted to reflect both positive and negative comments and provide as wide variety as possible. In some cases, we found that an explanation described a category in an illustrative way. Some of these explanations are presented as quotes. Observe that not all respondents explained their ratings. Slightly less than 50% of the respondents elaborated on their ratings. This means that the explanations offered might not give a complete picture. Still, many interesting issues were raised.

It is mostly straightforward to get help from the support system. The following descriptive terms were used in association with positive rankings: quick help (9 users), ease of use (5 users). On the negative side, the following descriptive terms were used: slow system (3 users), no help (2 users). One user giving a low score explained, "First wait for Java to start, then login, wait longer, enter a question, then (after waiting a bit) choose category, then read the FAQ and/or chat with assistant . . ."

Most users are willing to wait in a queue for human assistance but not for long. Several descriptive terms were used for unwillingness to wait long: impatience (8 users), online time is expensive (2 users), unreliable (2 users). Three users said that they were willing to wait for a long time because they could do many other things online while waiting.

Human assistants provide mostly good-quality support. The following descriptive terms were used in explanations: sensibility (10 users), flexibility (6 users), friendliness (6 users), making an effort to help (3 users), alternative solutions (1 user).

Textual chat is a viable means for help conversations on most topics. Several descriptive terms were used in explanations: simple (8 users), expressive (7 users), real time (2 users). On the negative side, the following terms were used: unstructured (3 users), inexpressive (3 users).

Human assistants make the use of a WIS more fun for most users. The following descriptive terms were used in explanations: human touch (12 users), making an effort to help (5 users), sense of humor (2 users). Three users thought the assistants had no effect on the fun of using a system. Two other users said that the system was too slow to have any effect on the fun.

Most users have a high level of trust in the advice provided by human Web assistants. The following descriptive terms were used in explanations: sensibility (14 users), advice is easy to verify (3 users). Two users commented that they generally wanted to verify pieces of advice before trusting it.

Human Web assistants have a positive influence on most users' trust in a WIS. The following descriptive terms were associated with this influence in the explanations: a message of caring (8 users), helpfulness (6 users), accountability (4 users). Two users said that they neither trust nor distrust a Website.

The presence of human Web assistants has a positive influence on the atmosphere of a WIS. The following descriptive terms were associated with this effect in explanations: alive and interactive (7 users), friendly (6 users), warm (4 users), personal (3 users), appealing and understandable (3 users). One user commented, "It's great! Sometimes I feel a bit alone when surfing the Web. But real people talking to you about your problems make you feel like you are part of a huge family. You are no longer alone out there." Another user said, "It gives a feeling of personal attention and makes the user feel important."

Most users are not put off by unavailability of assistants. However, some users are likely to never come back if they find the support system deserted. Their motivation is that they would believe that the system was not working, as is currently the case with many systems on the Web.

The support system can be used for different purposes. The support system does not only attract users who need help with a task that they cannot solve on their own. The system is also of value for more casual purposes, such as saving time or effort.

User Questionnaire 2

The results from the second user questionnaire are presented in Table 15.4. The percentages are calculated over the number of respondents (175). Some users gave more than one reason for not making use of a human assistant. Many users gave alternative reasons, according to option G. Most of these indicated that the users had registered out of need but then never got the time to actually try the support system during the 3-week period of the field study.

From the results, we note the importance of an easy-to-use support system. A user who cannot figure out how to use the support system is unlikely to maintain a

TABLE 15.4

Results From User Questionnaire 2

Reasons for Not Using a Human Assistant	
A. *The system didn't work.*	18 (10%)
B. *The FAQs solved the problem.*	26 (15%)
C. *There were no assistants logged in.*	50 (29%)
D. *I registered just to check things out.*	67 (38%)
E. *I didn't understand how to use the system.*	24 (14%)
F. *The system took too long to load.*	20 (11%)
G. *Other reasons*	24 (14%)

positive view of the WIS. The results also show that the FAQs system was useful in answering several users' questions.

Assistant Questionnaire

The following information is a summary of the most important issues that came out of the assistant questionnaire. The points that are raised follow issues in the open-ended questions. Notice that the presentation expresses the subjective opinions of the assistants, based on their experience during the field study. In our analysis of the questionnaires, we employed a method similar to the one we described for analyzing the statements in user questionnaire 1 above.

Users differ in conversational style and in background knowledge. The assistants need to be sensitive to these differences in order to adapt appropriately. For example, some users want quick answers to their questions, whereas others want to discuss alternative solutions. A few users can be impolite.

Users generally come straight to the point with their questions. Sometimes, however, follow-up questions are necessary to fully understand a user's problem.

Users explicitly express their opinion on the given advice. They generally tell the assistant what they think and whether they are satisfied or not.

An assistant does not need to speak English as a first language. Decent fluency and familiarity with the terminology of the domain seem to be enough. Observe, though, that this is likely dependent on the application domain. For example, with a banking application the situation could be different.

Being an assistant can cause stress. The following sources of stress were identified: hardware and software problems during help sessions, having to multi-task to find information for the user while chatting, having a slow or unreliable network, impolite users, and having difficulty expressing a solution in writing.

The most important skills or characteristics for an assistant are patience, domain knowledge, social skills, and fast typing. About one third of the responding assistants mentioned these skills. Several assistants also mentioned that it was important to be friendly, to have a sense of humor, and to be able to explain things in written words.

Textual chat is sufficient as a means of communication. Also, textual chat has an advantage over voice chat because the history of the chat is easily accessible in textual form, which means that the participants do not have to take notes while chatting. However, in some cases, the ability to graphically demonstrate things was missed in textual chat (mainly for art-related topics).

Handling more than one simultaneous help session can be difficult for some assistants. Eleven assistants thought they could handle several simultaneous help sessions, whereas 10 thought they could not. Most assistants agreed that their typing speed, users' conversation styles, and the kinds of questions affected their ability to handle simultaneous help sessions. Most assistants thought that three was the maximum number of possible simultaneous sessions.

LIMITATIONS

A field study is expected to bring out findings that can be generalized from the field setting to other similar real-world settings. The drawback is that there might be alternative explanations for the findings that were not eliminated in the design because of the generally uncontrolled nature of a field study. In this section we discuss such limitations and our approach to dealing with these.

Users in our study might have been loyal to the Elfwood site and thus only willing to give positive feedback. We discouraged such behavior by clearly pointing out that the support system was only a limited time experiment and that all kinds of feedback, whether positive or negative, were welcome. The assistants were volunteers with limited experience in helping others. Thus, they might not be representative of assistants with proper experience and education. We tried to make the best of the situation by providing instructions on how to handle the help conversations and by only selecting the volunteers who had proper skills within some area of expertise. We should also note that the participating users and assistants were all self-selected. Still, because the support system was operating in a real environment and the users used the system purely out of need and curiosity, the use likely reflects real use. A further limitation is that we cannot know if the persons who did not respond to the questionnaires differ systematically from the respondents.

It is reasonable that the results can be generalized to WISs similar to Elfwood. Such WISs include information providing sites such as Intranet systems and so-called Web presence sites ("sites that are marketing tools designed to reach consumers outside the firm," Isakowitz, Bieber, and Vitali, 1998, p. 78. Previously we carried out a proof-of-concept study on live help systems in an electronic commerce setting. The study gave indications similar to the findings reported here for user attitude (Aberg & Shahmehri, 2000). However, it is not obvious that all our recent results are valid for electronic commerce sites. For example, when it comes to trust, the situation for users is clearly different when trust in the advice might have financial consequences. Further studies should be carried out in electronic commerce contexts where users need to commit themselves to potentially risky decisions.

DESIGN IMPLICATIONS

Considering the data from the field study presented above, we can distinguish between two broad categories of findings. First, we have findings about the qualitative characteristics of live help systems and human assistants, such as users' attitude toward getting help from human assistants. Second, we have findings that have an impact on how live help systems should be designed. In this section, we investigate the design-related findings and focus our attention on the users' part of the live help system and their path from the realization of a need for help to the point where suitable help is provided by an assistant.

We have made a systematic analysis of the data collected from the two user questionnaires that aim at clarifying the implications for design. This analysis has been complemented by examples taken from a set of live help systems that exist on the Web today, roughly one year after the field study took place.[1] The considered systems were the first five working systems that we came across after browsing through the result from a Google[2] search on "live help system." Four of the systems were demonstration systems by companies selling live help systems, whereas the system in operation at the Galaxyphones Website was built using technology from LivePerson. These systems may be seen as representing the current commercial state of the art in live help. Now, we briefly go through the data collected from the user questionnaires, categorizing the design implications.

Considering the data from the questions presented in Table 15.1, we note that all findings regarding attitude are not directly design related. The data from questions 2 and 5, regarding flexibility and quality of support, did not yield any obvious design implications. The data from question 1, on the other hand, relates to the design of what might be called the start-up phase of the live help system, that is, the steps to be taken by the user until he or she gets connected to a human assistant. Questions 3 and 4 have implications for the design of a queuing system, whereas question 6 relates to the design of the communication between users and assistants.

Considering Table 15.2 and the reactions to the unavailability of assistants, we note that reactions A and B relate to the design for handling unavailability of assistants, although the implications are different. Reaction C relates to the design of the computer-based support and will not be considered here, whereas reaction D relates to the general design of the Website that is also beyond our current scope.

The data from the question in Table 15.3 about different usage types relates to qualitative characteristics of live help systems and does not have any direct implications for design.

Regarding Table 15.4 and the reasons for not using a human assistant, we note that reasons A and E are related to the general design of the live help system but not to any specific part of the system. Reason B is related to the design of the computer-based support and outside our current scope. Reason C relates to the unavailability of assistants, whereas reasons D and G have no apparent design implications. Finally, reason F clearly relates to the start-up phase of the live help system.

In summary, the field study has resulted in a number of implications for the design of live help systems that can roughly be categorized as follows:

---

[1]The systems considered were the LiveContact system by ServiceSoft, the WebSiteHelp system, the SiteAssistant system, the LivePerson system, and the live help system at the Galaxyphones Website.

[2]Web search engine, available at http://www.google.com

- Start-up phase.
- Unavailability of assistants.
- Queuing.
- Communication.

Next, we discuss these implications in more detail before providing concrete design guidelines.

Issues

Start-up Phase

As mentioned previously, many Elfwood regulars failed to notice the presence of the live help system during the field study period. This very first step toward a consultation dialogue with a human assistant is of obvious importance. The user must be aware of the existence of the live help system. This observation leads us to the first three design issues:

1. How to make users aware of the live help service? Unless live help becomes a standard at most Websites, it is important to emphasize the availability of live help. Although most users thought it was straightforward to get help, some users pointed out that the process for getting human help was too slow (see the explanations to question 1). This was partly due to the many steps the users had to go through before reaching a human assistant. Minimizing the time spent on these steps would thus be a good thing from the user's point of view. One important design issue then relates to the registration and log-in steps that were used for our system.

2. Should there be some form of registration and log-in procedure for users, and in that case, what kind of user information should be collected?

Another issue related to the time spent on the start-up phase concerns the implementation technology. In order to minimize the time spent, the fastest solution is preferable.

3. Should a live help system rely on client-side or server-side computation?

Unavailability of Assistants

We reported that most users are not put off by the unavailability of human assistants and that many users would simply try again at a later time (see the explanations to the question in Table 15.2). We also reported that a large share of the users who did not connect to a human assistant did so because of unavailability of assistants (see Table 15.4, reason C). The question that all these users must consider then is when to try again. Also, a few users in our field study were apparently put off by this kind of unavailability and would never return again (see reaction B, Table 15.2). Hence, we consider the fourth design issue:

4. Unavailability of assistants. Unless the Website makes sure to have human assistants online at all times, the system must deal with the unavailability of assistants. The design problem is how to inform the users about the unavailability situation so that they are not put off and so that their attempts to get help are facilitated.

Queuing

From the field study we found that most users are willing to wait in a queue for human assistants, but not for long. The reasons for not being willing to wait long relate to impatience, unreliability of Web systems in general, and expensive online time (see questions 3 and 4, and their explanations). The only way to deal with the latter reason seems to be to minimize the waiting time. The first two reasons have important implications for the design of a queue system. It is important to keep the user informed about what is going on behind the scenes of the queuing system:

5. Status information for queuing. A user who has to wait in a queue before getting live help may need some kind of status information.

Considering that some users are likely to leave the queue if they have to wait for too long (no matter what kind of status information is presented), it is important that these users do not leave with a bad impression of the Website. During our field study, we observed several cases in which users waited for a long time and then left and never came back to the help system (see also reaction B, Table 15.2). This is our next design issue:

6. How to deal with the situation where a user decides to leave the queue?

Some users said that they were indeed willing to wait in a queue for a long time, simply because they could do other things at their computer while waiting (see the explanations to questions 3 and 4). This brings up another design issue:

7. How can a queuing user be alerted when it is his or her turn to get help?

Communication

Most users found text chat to be a viable communication means for consultation dialogues. However, some users found text chat to be unstructured and inexpressive (see the explanations to question 6). Thus, depending on the Website domain, text chat alone may not be a sufficient means of communication:

8. Is text chat sufficient for live help, or is there a need for complementary or alternative means of communication?

Design Guidelines

Guideline 1

Highlight the presence of a live help system on the Website. Until live help systems become a standard feature on Websites in general, it is necessary to emphasize the existence of such support so that users become aware of it. This is especially important for users who are already familiar with the Website and may fail to observe modifications to the site unless they are clearly advertised. The live help system at Elfwood was advertised by a large icon prominently displayed on the home page. We also had smaller icons added to every single page at Elfwood. Still, after our field study, we heard from several Elfwood members that they had completely failed to notice the existence of the live help system at the site.

The live help system at the Galaxyphones Website is advertised through a button with flashing text placed in a prominent position on the top left part of the site. Because the button is often the only thing flashing, most users will notice it quickly. We do not recommend the use of such a technique in the long run because it would only distract the users from the actual site content, but it can be useful in an initial phase, just after the live help system has been introduced. Also, if the site has some form of session management, the flashing effect can be turned off a few seconds after the user entered the site or even suppressed completely for users who have already seen it.

Guideline 2

Simplify the initiation process for live help as much as possible. Requiring the user to go through a lengthy registration process, where lots of demographic data has to be filled in, should be avoided if possible. If information about the user is important for the Website, it can be obtained as part of the dialogue between the user and an assistant instead. In our own system, we admittedly made use of a registration phase with a log-in name and a password. But the reason for this was that we wanted to make sure that we had the e-mail address of all users so that we could send out questionnaires after the system deployment period. Unfortunately, this may have turned many potential users off. The live help system at the LivePerson Website does not require the user to provide any personal information, making the initiation process very smooth. On the other hand, the ServiceSoft LiveContact system requires the user to type in a first and last name and an e-mail address before the live help can start, thus delaying the help.

Guideline 3

The user's interface of our live help system was implemented as a Java applet. According to our experiences, this is something to be avoided. Different versions of different Web browsers run different Java implementations. This means that it is very difficult to make the system behave well on all possible computer platforms.

Even though we restricted our Java code to Java version 1.0, which was supposed to be supported by all modern versions of Netscape and Microsoft Internet Explorer, we had major problems with some browser versions. In addition to this, the rather long loading and starting time for the applets annoyed several users, especially those with slow Internet connections. Consequently, we suggest that Java applet solutions be avoided in favor of server-side solutions whenever technically possible. Still, whatever technical solution is chosen, testing on most common browsers and operating systems is necessary. As an example, the WebSiteHelp system worked fine with Internet Explorer on Unix but did not work with Netscape.

Guideline 4

Make availability hours clear to the users. The times and days when human assistants are available should be clearly presented. Also, at those times when there are no assistants available and a user tries to use the live help system, the user should be informed about the unavailability and when human assistants will be available again. For Websites with an international user base, it is important to give the availability times relative the local time zone. Preferably, there should be a feature allowing the user to display the availability times according to his or her local time zone. The SiteAssistant system simply gave the users a message along the lines of "try later" when there were no assistants available. This should definitely be avoided. The SiteAssistant system did, however, allow the user to formulate a question and submit his or her contact information so that an assistant could get back to the user at a later time.

Guideline 5

Provide a queuing user with as much queue status information as possible. The information should include current queue position, expected remaining waiting time (see, e.g., Foster & De Reyt, 1999), and other kinds of relevant awareness information about the assistants. The information should be updated frequently. It is not necessary to clutter the queuing display with all information at once, but it should be accessible to the user.

In our field study system, we implemented only queue position, as illustrated in Fig.15.3. In our system, there are different queues for different question types. This is information that the user need not know about and can indeed be a source of confusion. We also provided the user with the option of changing to a queue with an available assistant, if possible. In practice, this feature led to many confusing situations for both users and assistants. Typically, a user who clicked on the button for changing queue ended up getting help by an assistant who had no knowledge in the topic of the question. The idea of having a "fast queue" for quick questions of a general nature is interesting, but it must be made clear to the user how it works and what the rules are. In the LiveContact system by ServiceSoft, the queuing user is updated by a scrolling information text. This can be a good means for keeping

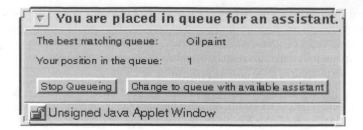

FIG. 15.3. Screen shot from the queue display.

the user informed of what is happening, but it is important that the text does not repeat itself.

Guideline 6

A user who decides to leave the queue should be given the option to have the live help system contact him or her at a later time. This way the waiting time may not feel like a complete waste of time for the user, and negative attitudes towards the site may be avoided. Such a function could be implemented by automatically providing the user with a form where the question can be formulated and sent to the system. The live help system can then get back to the user through e-mail or through regular telephone, depending on the user's preference. This feature is nicely implemented in the LiveContact system by ServiceSoft.

Guideline 7

A user who is next in line for live help must be alerted both visually and via audio. Our results showed that some users are willing to wait for help in a queue is because they can do other things on their computers while waiting. This implies that the queue display may not be under visual observation continuously. Thus, an audio signal could be useful to alert the user. It is also important to inform the users of how the alerting is done so that they are aware that other things can be done at the same time.

Guideline 8

Provide alternative means of communication. While text chat was found to be a suitable means of communication by most users, some users may prefer less personal communication, for example, through e-mail. Technically, adding e-mail to the live help system is not much of a problem. Another option is to allow voice communication, for example, through the telephone. This could be implemented so that the user submits his or her phone number and a time frame for being called back. An assistant of the live help system then initiates the communication by calling the user. A problem with this approach is the potentially large cost for the Website. Another problem is that many users connect to the Internet through their telephone connection. Thus, they cannot be connected to the Website while communicating with an assistant. Finally, there may be domains where text or even voice is not a sufficient means for communication. This could be the case for domains with a large dependence on visualization. This was the case for our field study, where many of the consultation dialogues were about art-related issues. In cases like this, special means of communication may need to be supported. For example, a function for quickly exchanging images as part of the communication might be a useful extension to regular text chat.

RELATED WORK

Since the initiation of our work, there have been commercial moves in the directions discussed in this chapter. Companies such as LivePerson and FaceTime now offer commercial live help systems. This trend increases the importance of the kind of studies reported in this paper. While the efforts by FaceTime and LivePerson are well motivated, little is known about how users interact with these kinds of systems and how the most benefit can be derived from the human assistants.

The systems currently offered by LivePerson and FaceTime are clearly similar in spirit to the system we have implemented. Still, there are some important differences. First, user modeling seems to play a limited part in the commercial systems when compared to our system. Second, the user support provided is completely centered on the human support agents. Our approach, routing users through a computer-based question-answering system before connecting users with human assistants, seems to be a better use of resources (Aberg & Shahmehri, 2001a).

In Kobayashi et al. (1998), two applications for Web-based collaborative customer service are described. The applications are for a banking kiosk setting and for home banking over the Internet. The banking kiosk application uses a shared browser approach with ink annotation for communication. The home banking application also uses a shared browser approach and has support for voice chat. The two applications are related to our approach. They are, however, limited when it comes to providing personalized support. The descriptions of the systems are on an architectural level, and no evaluation is presented.

In Losleben, Tull, and Zhou (1997), the Help-Exchange system is described. The general idea behind the system is that all persons have some kind of expertise that they may share with others in exchange for help on topics where they are not experts. The system uses a hierarchical structure for different areas of expertise. A new user can sign up as an expert for one or more areas. The system can then assign the user questions. The user also has the right to ask a number of questions that other users with different kinds of expertise can answer. The number of questions that a user is allowed to ask in the system is related to the number of questions he or she has answered.

The Answer Garden (AG) system by Ackerman (1994) is a question-answering system similar to Help-Exchange and thus related to our work. Still, these two systems have some important differences to our live help system. Neither of the systems support personalization similar to our user-modeling component. Further, where our system supports synchronous communication between users and assistants via textual chat, question answering by an expert in AG and the Help-Exchange system consists of two asynchronous messages, the question and the answer.

The AG system was later extended to a second version called Answer Garden 2 (AG2; Ackerman & McDonald, 1996). AG2 features mainly two new functions consisting of an automatic escalation of help sources and a form of collaborative refinement of the system database taking sources other than user questions into consideration. These features could very well be incorporated into our system and would be interesting subjects for future work.

ADDITIONAL ISSUES

Achieving universal usability is a very well-motivated although difficult objective. From our findings, it is clear that humans have an important role to play in user support for WISs and, in the long run, for achieving universal usability. However, we are just at the beginning of this research, and there are ample opportunities for further interesting investigations.

We have shown that human assistants have a positive influence on user attitudes toward a Website. Still, the direct influence of these attitude variables needs to be studied, for example, on site loyalty for different kinds of WISs such as Web presence sites or electronic commerce sites. Although it is suggested that trust has a positive influence on customer loyalty for electronic commerce sites (Lee et al., 2000), there are still many open questions.

Trust is a complex concept (Ratnasingham, 1998), and it is difficult to interpret our results on trust because of the simple nature of the questionnaire statements. Future work should more thoroughly investigate the effect of human assistants on users' trust in a WIS. Another concept worth studying is playfulness, which can have a positive influence on a user's learning, mood, involvement, and satisfaction (Webster & Martocchio, 1992). Do human assistants evoke users' playfulness?

The 3-week duration of our field study is another limitation. It is possible that usage patterns would change over a longer time period. Further work is needed to study the longitudinal effects on this kind of user support system.

The study by Lohse and Spiller (1998) suggests that customer service influences neither site traffic nor sales. However, the variables used in the study to represent customer service are limited (e.g., presence of a FAQs section, e-mail support, or a phone number to a customer service department). Thus, the results cannot be generalized to more advanced forms of customer service, and further studies are needed to see what effects the form of customer service that we propose in this paper would have on the variables mentioned.

In Jensen, Farnham, Drucker, and Kollock (2000), the effect of communication modality on trust and cooperation in online environments is studied. The results do not indicate a statistically significant difference between "no communication" and communication via textual chat, whereas voice chat has a significantly better effect on trust and cooperation. This indicates that users' trust for the support of human assistants could be further increased if voice communication were to be used instead of textual chat.

In Rocco (1998), collaborative tasks requiring trust are studied. It is shown that trust is difficult to establish if communication is carried out electronically. It is also shown that initial face-to-face contact between people can help in establishing trust when communication is also done electronically. Although the users in our study did not risk much by trusting the assistants, it may be a different situation, for example, in banking applications. It would be interesting to study if there are alternative means to initial face-to-face contact for maintaining trust in electronic contexts.

CONCLUSION

The live help system is a new and fascinating concept, building on the idea of integrating humans with computer-based functions. This idea may have been around since the beginning of artificial intelligence, but it has not been exploited to any large extent. Today, in the networked world, it is getting increasingly feasible to integrate humans into the run-time behavior of computer systems. Live help systems are but one possible application. One could easily imagine other applications that would potentially benefit greatly from having a human–computer collaboration for problem solving. Now, as live help systems begin to gain in popularity (both among researchers and practitioners), we hope that the collaboration between computer functions and humans will receive greater attention among researchers in computer science and related areas.

REFERENCES

Aberg, J., & Shahmehri, N. (2000). The role of human Web assistants in e-commerce: an analysis and a usability study. *Internet Research: Electronic Networking Applications and Policy, 10*(2), 114–125.

Aberg, J., & Shahmehri, N. (2001a). Collection and Exploitation of Expert Knowledge in Web Assistant Systems. In Ralph H. Sprague, Jr. (Ed.), *Proceedings of the 34th Hawaii International Conference on System Sciences*. Maui, Hawaii, USA: IEEE.

Aberg, J., & Shahmehri, N. (2001b). An Empirical Study of Human Web Assistants: Implications for User Support for Web Information Systems. In Julie A. Jacko, Andrew Sears, Michel Beaudouin-Lafon, & Robert J. K. Jacob (Eds.), *Proceedings of the CHI Conference on Human Factors in Computing Systems* (pp. 404–411). Seattle, WA: Association for Computing Machinery.

Aberg, J., Shahmehri, N., & Maciuszek, D. (2001). User Modelling for Live Help Systems. In Ludger Fiege, Gero Mühl, & Uwe G. Wilhelm (Eds.), *Proceedings of the Second International Workshop on Electronic Commerce* (pp. 164–179). Heidelberg, Germany: Springer.

Ackerman, M. S. (1994). Augmenting the Organizational Memory: A Field Study of Answer Garden. In *Proceedings of CSCW '94* (pp. 243–252). Chapel Hill, NC, USA: Association for Computing Machinery.

Ackerman, M. S., & McDonald, D. W. (1996). Answer Garden 2: Merging Organizational Memory with Collaborative Help. In M. Acherman (Ed.), *Proceedings of CSCW '96* (pp. 97–105). Boston, MA, USA: Association for Computing Machinery.

Bouchard, T. J., Jr. (1976). Field research methods: Interviewing, questionnaires, participant observation, systematic observation, unobtrusive measures. In M. D. Dunnette (Ed.), *Handbook of industrial and organizational psychology* (pp. 363–413). Chicago: Rand McNally.

Carroll, J. M., & McKendree, J. (1987). Interface Design Issues for Advice-Giving Expert Systems. *Communications of the ACM, 30*(1), 14–31.

Fogg, B. J., Marshall, J., Laraki, O., Osipovich, A., Varma, C., Fang, N., Paul, J., Rangnekar, A., Shon, J., Swani, P., & Treinen, M. (2001). What Makes Web Sites Credible? A Report on a Large Scale Quantitative Study. In Julie A. Jacko, Andrew Sears, Michel Beaudouin-Lafon, & Robert J. K. Jacob (Eds.), *Proceedings of the CHI Conference on Human Factors in Computing Systems* (pp. 61–68). Seattle, WA: Association for Computing Machinery.

Foster, R. H., & De Reyt, S. (1999). Re-inventing the Call Centre with Predictive and Adaptive Execution. *British Telecommunications Engineering, 18*, 180–184.

Hoffman, D. L., Novak, T. P., & Peralta, M. (1999). Building Consumer Trust Online. *Communications of the ACM, 42*(4), 80–85.

Isakowitz, T., Bieber, M., & Vitali, F. (1998). Web Information Systems. *Communications of the ACM, 41*(7), 78–80.

Jarvenpaa, S. L., & Todd, P. A. (1997). Consumer Reactions to Electronic Shopping on the World Wide Web. *International Journal of Electronic Commerce, 1*(2), 59–88.

Jensen, C., Farnham, S. D., Drucker, S. M., & Kollock, P. (2000). The Effect of Communication Modality on Cooperation in Online Environments. In T. Turner, G. Szwillus, M. Czerwinski, & F. Paterno (Eds.), *Proceedings of CHI '00* (pp. 470–477). The Hague, The Netherlands: Association for Computing Machinery.

Karat, J. (1997). User-Centered Software Evaluation Methodologies. In M. Helander, P. Landauer, & P. Prabhu (Eds.), *Handbook of human–computer interaction* (pp. 689–704). Amsterdam: Elsevier.

Kobayashi, M., Shinozaki, M., Sakairi, T., Touma, M., Daijavad, S., & Wolf, C. (1998). Collaborative Customer Services Using Synchronous Web Browser Sharing. In *Proceedings of CSCW '98* (pp. 99–108). Seattle, WA, USA: Association for Computing Machinery.

Kraut, R., Scherlis, W., Mukhopadhyay, T., Manning, J., & Kiesler, S. (1996). The Home Net Field Trial of Residential Internet Services. *Communications of the ACM, 39*(12), 55–63.

Lee, J., Kim, J., & Moon, J. Y. (2000). What makes Internet Users visit Cyber Stores again? Key design factors for customer loyalty. In T. Turner, G. Szwillius, M. Czerwinski, & F. Paterno (Eds.), *Proceedings of CHI '00* (pp. 305–312). The Hague, The Netherlands: Association for Computing Machinery.

Lohse, G. L., & Spiller, P. (1998). Quantifying the Effect of User Interface Design Features on Cyberstore Traffic and Sales. In C.-M. Karat, A. Lund, J. Coutaz, & J. Karat (Eds.), *Proceedings of CHI '98* (pp. 211–218). Los Angeles, CA, USA: Association for Computing Machinery.

Losleben, P., Tull, A. S., & Zhou, E. (1997). Help Exchange: an arbitrated system for a help network. *Journal of Network and Computer Applications, 20*, 123–134.

Löwgren, J. (1993). *Human–computer interaction*. Lund, Sweden: Studentlitteratur.

Ratnasingham, P. (1998). The importance of trust in electronic commerce. *Internet Research: Electronic Networking Applications and Policy, 8*(4), 313–321.

Rocco, E. (1998). Trust breaks Down in Electronic Contexts but Can Be Repaired by Some Initial Face-to-Face Contact. In C.-M. Karat, A. Lund, J. Coutaz, & J. Karat (Eds.), *Proceedings of CHI '98* (pp. 496–502). Los Angeles, CA, USA: Association for Computing Machinery.

Shneiderman, B. (2000). Universal Usability. *Communications of the ACM, 43*(5), 85–91.

Spiller, P., & Lohse, G. L. (1998). A Classification of Internet Retail Stores. *International Journal of Electronic Commerce, 2*(2), 29–56.

Vijaysarathy, L. R., & Jones, J. M. (2000). Print and Internet catalog shopping. *Internet Research: Electronic Networking Applications and Policy, 10*(3), 191–202.

Webster, J., & Martocchio, J. J. (1992). Microcomputer playfulness: Development of a measure with workplace implications. *MIS Quarterly, 16*(2), 201–226.

16

Trust in Cyberspace

Brian P. Bailey
Laura J. Gurak
Joseph A. Konstan
University of Minnesota

Trust plays a critical role when a user assesses the believability of online information or when selecting an exchange site from which to purchase a product. Users will not believe or participate in a transaction with those whom they do not trust. When a design team develops an informational or exchange site, it is responsible for ensuring that a user perceives that site as trustworthy. Thus, the goal of this chapter is to provide Web designers with a definition of trust, an understanding of where trust originates and how it is perceived by a user, and a mechanism in the form of a trust taxonomy for analyzing the trust production methods of an exchange site.

INTRODUCTION

The Internet provides a user with access to a myriad of online stores, selling products ranging from books and computers to automobiles and houses. But when a user is ready to purchase a product online, how does that user assess which online stores are trustworthy and which are not? In the brick-and-mortar world, a user can rely on physical cues such as the neighborhood location, physical size, presence of customers, and interior décor of a store to help assess the store's trustworthiness. However, when assessing the trustworthiness of an online store, those same physical cues are not available, and a user must rely on other cues such as the privacy policy, visual aesthetics, and navigation quality of the online store.

Although the principles of trust presented in this chapter help explain how a user assesses the believability of information found on news groups, discussion forums, and informational Websites, this chapter focuses on trust as it relates to computer-mediated exchange. The term *computer-mediated exchange* (CME) is used to collectively refer to the business-to-consumer and consumer-to-consumer electronic exchange models. For example, a consumer purchasing antiques, books, or herbal remedies from eBay, Amazon.com, or Drugstore.com qualifies as a CME. The terms *exchange partner* and *exchange site* are used to refer to the person or Website involved in a CME, respectively.

Working to develop a trustworthy exchange site is important for several reasons. Trust is required for all willing transactions and, without it, no market could function (Zucker, 1986). Trust creates more favorable attitudes toward suppliers as well as customer loyalties (Schurr & Ozanne, 1985) and "helps partners project their exchange relationships into the future" (Doney & Cannon, 1997). Trust enhances competitiveness, reduces transaction costs, and mitigates opportunism in uncertain contexts (Doney & Cannon, 1997). In sum, working to develop a trustworthy exchange site yields a competitive advantage.

When a design team develops an exchange site, it is responsible for ensuring that a user perceives the site as trustworthy. However, ensuring trustworthiness is not the same as ensuring usability. Adding a privacy policy, testimonial, or accreditation to an exchange site ostensibly has no impact on a user's task performance but may have a large impact on the user's perception of the site's trustworthiness. Thus, the goal of this chapter is to provide Web designers with a definition of trust, an understanding of where trust originates (sources of trust) and how trust is perceived by a user (dimensions of trust), and a mechanism in the form of a trust taxonomy for analyzing the trust production methods of an exchange site.

DEFINITION OF TRUST AND TRUSTWORTHINESS

Defining trust-related terms is most effective when considering the entire process of trust production. The trust production process begins with a trustee possessing an objective, intrinsic level of trustworthiness. That is, the degree to which the trustee will fulfill the transactional obligations during an exchange is known to the trustee. However, because a trustor cannot precisely know that intrinsic value, the trustor must perceive extrinsic cues produced from the trustee in order to attribute a level of trustworthiness. Thus, we define *trust* as the perception of the degree to

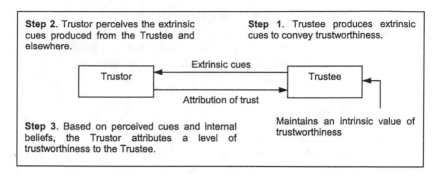

FIG. 16.1. The trust production process.

which an exchange partner will fulfill the transactional obligations in situations characterized by risk or uncertainty. However, trustworthiness must be defined from both the perspective of the trustee and the trustor. From the perspective of the trustee, we define *trustworthiness* as an objective quality governing the degree to which transactional obligations will be fulfilled in situations characterized by risk or uncertainty. From the perspective of the trustor, we define *trustworthiness* as an attribution of trust. The trust production process is summarized in Fig. 16.1.

The attribution of trust occurring in Step 3 of Fig. 16.1 may be greater than, less than, or equal to the trustee's intrinsic value of trustworthiness, each case having a significant implication for the exchange (Brainov & Sandholm, 1999).

The trust production process just described is reasonable regardless of whether the trustee is another person or a technological entity such as a Website. Substantial evidence exists indicating that humans interact with technology in a social manner (Nass, Moon, Fogg, Reeves, & Dryer, 1995; Nass, Reeves, & Leshner, 1996; Nass, Steuer, & Tauber, 1994; Waern & Ramberg, 1996). That is, when humans perceive specific social cues produced by a technological entity, their reactions are social reactions. Further, evidence exists indicating that people develop trust not only in salespeople, but also in their suppliers (Doney & Cannon, 1997). Together, these results confirm that humans develop trust in an exchange site in the same manner that they develop trust in other people.

SOURCES AND DIMENSIONS OF TRUST

In the trust production process, extrinsic trust cues are produced from multiple sources and perceived along several dimensions leading to an attribution of trust. A *trust source* refers to the belief, impression, experience, or institution from which a trust cue is produced. A *trust dimension* is an operational attribute of trust to which a trust cue contributes. A person cognitively combines the trust dimensions forming an overall attribution of trust. The sources and dimensions of trust are related in a many-many relationship; that is, a single trust source may contribute to multiple dimensions of trust, whereas multiple sources may contribute to a single dimension of trust. The next two sections provide a closer look at these two concepts.

Sources of Trust

From a synthesis of relevant literature in human–computer interaction (Fogg & Tseng, 1999; Kim & Moon, 1997; Tseng & Fogg, 1999), social psychology (Deutsch, 1973; Giffin, 1967; Patzer, 1983; Rempel, Holmes, & Zanna, 1985; Rotter, 1980), marketing (Doney & Cannon, 1997; Johnson & Grayson, 2000), and economics (Zucker, 1986), the four sources of trust are:

• **Presumptions.** Presumptions produce trust through general beliefs or levels of confidence maintained in the absence of doubt. These beliefs and confidence levels are derived from general assumptions and stereotypes existing within one's own culture. For example, an exchange partner may presume that an exchange site is less trustworthy than a brick-and-mortar retailer.
• **Surface inspection.** Surface inspection produces trust through an examination of an exchange partner's external appearance, such as the visual design of a Website or the physical appearance of a person. Once formed, first impressions can be extremely difficult to break (Zajonc, 1980).
• **Experience.** Experience produces trust through repeated successful exchanges with an exchange partner. For example, a person having repeated successful exchanges with an exchange partner will likely perceive that partner as being more trustworthy than an unfamiliar partner. Reputations and brands can also be considered as experience-based sources of trust (Zucker, 1986).
• **Institutions.** Institutions or third parties produce trust through what they report about an exchange partner, that is, whether an exchange partner deserves the "Good Housekeeping Seal of Approval" or not. In this case, the production of trust is a transfer of trust from the institution to the exchange partner, where the amount of transference is proportional to the perceived trustworthiness of the institution. Although the name institution implies a governing body, this source of trust also includes recommendations from family members, friends, or colleagues.

The sources of trust are not mutually exclusive, and the contribution of each may change over time. For example, in the absence of previous exchanges and third party recommendations, an exchange partner must rely on presumptions and surface inspection to make an initial attribution of trust. In a future exchange with that same partner, this past experience will contribute more whereas presumptions and surface inspection will contribute less.

Dimensions of Trust

Although trust cues originate from sources of trust, dimensions of trust are the operational attributes to which they contribute. Drawing mainly from the social psychology literature (Giffin, 1967; Patzer, 1983; Posner & Kouzes, 1988; Rempel, Holmes, & Zanna, 1985), the seven dimensions of trust are:

- **Attraction of an exchange partner's physical or nonphysical characteristics.** For example, an attractive person is generally perceived as being more trustworthy than an unattractive person (Patzer, 1983). Analogously, an aesthetically pleasing exchange site should be perceived as more trustworthy than one that is not.
- **Dynamism of the additional (peripheral) communication provided by an exchange partner through oral, written, or visual communication channels.** For example, a salesman's body language or a Website's ticker-tape display can be regarded as dynamism.
- **Expertness of an exchange partner's relevant skill, ability, or knowledge.** An expert is generally perceived as being more trustworthy than a nonexpert (Brainov & Sandholm, 1999; Peters, Covello, & McCallum, 1997).
- **Faith that an exchange partner will fulfill obligations despite an uncertain future.** Faith is important in exchanges where past experiences are minimal or only indirectly related to the current exchange. For example, suppose a buyer who has previously purchased several inexpensive items from a seller is now considering the purchase of an expensive item from that same seller. The buyer must have faith that the seller will once again act responsibly and fulfill any obligations.
- **Intentions of an exchange partner in terms of perceived goals and objectives.** For example, a seller who is open, honest, and discloses relevant information is perceived as having a genuine interest in the welfare of the buyer and, thus, is perceived as being more trustworthy (Doney & Cannon, 1997).
- **Localness of an exchange partner's ideals, beliefs, values, or geography.** The closer the perceived proximity along one or more of these dimensions, the more trustworthy an exchange partner appears. For example, a buyer may perceive a seller who donates to or volunteers for the same charitable organization as being more trustworthy than a seller who does not.
- **Reliability of an exchange partner measured in terms of dependability, predictability, or consistency.** Reliability is directly rooted in past experiences and prior interactions. For example, a buyer who has successfully purchased books from a seller will perceive that seller as being more trustworthy than an unknown seller.

The relationship between the sources and dimensions of trust is depicted in Fig. 16.2. By combining the sources and dimensions of trust into a matrix, with the sources as rows and the dimensions as columns, a trust taxonomy is created.

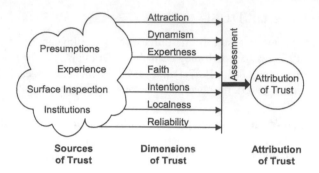

FIG. 16.2. An expanded view of Steps 1 and 2 of the trust production process shown in Fig. 16.1. Trust cues are produced from one or more sources, perceived along one or more dimensions, and cognitively assessed by a user, resulting in an overall attribution of trust.

This trust taxonomy, or classification matrix, provides a concrete mechanism for performing trust analysis of an exchange site. Trust analysis of an exchange site is performed by first inspecting that site for distinct methods of trust production and then placing each method into one or more of the matrix cells.

CASE STUDY—TRUST PRODUCTION METHODS IN CME

In this case study, we first identify the primary trust production methods used within Drugstore.com and then classify those methods according to our trust taxonomy. Drugstore.com (www.drugstore.com) is a leading online retailer of health, beauty, and pharmacy products. Inspection of the site reveals that Drugstore.com's trust production methods include:

• **Affiliations with companies who display links back to Drugstore.com.** The quantity, quality, and diversity of third party sites displaying links back to Drugstore.com helps to produce a sense of localness and reliability.

• **Brand of the company (E-commerce trust study, 2000; Schurr & Ozanne, 1985) as well as the products it offers.** Brands convey a specific image, reputation, and presumed quality of service that help produce a sense of expertness, localness, and reliability to a buyer.

• **Charitable organizations supported by the company.** By informing a customer about its charitable donations, Drugstore.com helps produce a sense of good intentions and localness.

• **Customer service.** The site provides extensive help and support for order processing, payment, shipping, tracking, account management, prescription and insurance issues, and FAQs. This extensive customer service section produces a sense of good intentions, availability (localness), reliability, and future dependability.

• **Functionality, or "feel" of the site.** This is characterized by working links, the speed of page loading, and intuitive nature of order processing (E-commerce trust study, 2000). Drugstore.com maintains a high degree of functionality, which produces a sense of expertness and reliability.

• **Investor relations links.** These offer a complete suite of investor information, such as Securities and Exchange Commission (SEC) filings, stock quotes and charts, and earnings estimates. By supplying this information, Drugstore.com produces a sense of reliability and good intentions.

• **Navigation of product structure.** This is measured in terms of accuracy, efficiency, and search methodology (E-commerce trust study, 2000). By enabling a user to browse products by brand or category or search for a specific product, Drugstore.com produces a sense of helpfulness (good intentions), reliability, and expertness.

• **Presentation of the site.** This is determined by its visual appearance and quality of content structure (E-commerce trust study, 2000). Quality presentation produces a sense of attraction and expertness, resulting in an increased perception of trustworthiness (Kim & Moon, 1997).

• **Privacy policy.** The site should clearly explain if, when, why, and under what circumstances a customer's personal information, for example, e-mail address, home address, or credit card number is needed, used, stored, or traded to third parties. The privacy statement also explains how to update stored personal information as well as explains the use of cookies and encryption. By disclosing this information, Drugstore.com draws on our cultural beliefs that those who are open and honest have good intentions both now and in the future (Doney & Cannon, 1997).

• **Relationship building (Doney & Cannon, 1997; E-commerce trust study, 2000).** This is done by providing a customer with personalized e-mail notifications, describing new product offerings, press releases, and medical alerts, and welcoming a recognized customer to the site. By working to build a relationship with a customer, Drugstore.com produces a sense of good intentions, faith, and localness for that customer.

• **Seals of approval (E-commerce trust study, 2000; Schurr & Ozanne, 1985).** These depict privacy compliances such as TrustE, business standards such as BBBonline, or specialized certifications such as Verified Internet Pharmacy Practice Site. By including these symbols, Drugstore.com produces a sense of expertness, good intentions, and reliability in proportion to a customer's perceived trustworthiness and selectivity of these third party seals of approval.

• **Size and market share (Doney & Cannon, 1997).** Size and marketshare conveys that many other customers trust Drugstore.com enough to do business with them. On its home page, Drugstore.com lists how many packages have been

TABLE 16.1
Classification of Trust Production Methods for Drugstore.com

	Attraction	Dynamism	Expertness	Intentions	Faith	Localness	Reliability
Presumed			B	Ch, Pi	Pi	B, Ch	B
Surface	Pe	Si	Pe	Ch, Cu, I, Si	Cu, I	Cu	Cu, I, Si
Experience		R	B, F, N, Pe	N, R	R	B, R	B, F, N
Institutional			Se, T	Se, T	T	A, T	A, Se, T

A = Affiliations, B = Brand, Ch = Charitable organizations, Cu = Customer service, F = Functionality, I = Investor relations, N = Navigation, Pe = Presentation, Pi = Privacy policy, R = Relationship building, Se = Seals of approval, Si = Size and marketshare, and T = Testimonials.

shipped and which product was recently sold. Working to convey size and market share produces a sense of reliability and good intentions.

• **Testimonials from previous customers and the popular press.** By sharing these positive statements, Drugstore.com produces a sense of expertness, good intentions, localness, faith, and reliability to a customer.

These methods of trust production have been classified according to our trust taxonomy in Table 16.1.

Inspection of the table reveals several interesting features. First, Drugstore.com's trust production methods draw heavily on direct experience to strengthen many of the trust dimensions. As a result, a customer should be encouraged to explore Drugstore.com's features in order for attribution of trust to meaningfully increase, and methods for encouraging exploration should be analyzed during usability testing. Second, Drugstore.com effectively leverages the interactive and dynamic nature of the Internet to convey a sense of trustworthiness through the dimensions of expertness, good intentions, and reliability, using trust cues such as seals of approval, charitable organization support, and extensive customer service.

DISCUSSION

Using the trust taxonomy in this case study and others (Bailey, Gurak, & Konstan, 2001) has produced two important lessons:

• **The same person(s) should perform all classifications of trust production methods to maintain consistency and reliability.** The classification process, known as *coding*, is subjective rather than an application of hard and fast rules. For example, because a first-time user can only inspect the links for customer service, we classified it as a surface-based source of trust, although the actual quality of that service is something that the user must experience. However, overcoming the challenges of subjective coding practices is not insurmountable, as analogous situations have long been addressed in the area of communication research (Rice & Love, 1987; Sproull & Kiesler, 1986; Walther, 1992).

- **Using the trust taxonomy for trust analysis works.** Even though the exchange site chosen for the case study is mature and presumably already perceived as trustworthy by most, trust analysis has identified some missed opportunities for trust production. For example, Drugstore.com could add to the dimensions of attraction and localness by providing a welcome letter from the CEO or add to the dimension of dynamism through a community support program.

The primary limitation of the taxonomy is that the exact contribution that each matrix cell or trust production method makes to an overall attribution of trust has not been determined. As a result, a design team must rely on the density of the matrix coupled with an intuitive understanding of the trust production methods in order to estimate a user's attribution of trust. However, other research has demonstrated that an attribution of trust can be empirically measured (Deutsch, 1958; Kim & Moon, 1997; Rocco, 1998), including a subset of the trust dimensions presented in this chapter (Doney & Cannon, 1997; Giffin, 1967). Providing a more quantitative approach to assessing the trustworthiness of an exchange site will be a challenging task for future research.

Finally, trust is a dynamic process that strengthens or weakens over time, based on a customer's experience. Thus, an exchange site must provide a strong set of trust cues to establish an initial perception of trust, but afterward, that site must provide positive experiences for a customer in order to strengthen—or at least maintain—that initial perception of trust.

CONCLUSION

Creating a trustworthy exchange site is a critical requirement for Web designers, as online consumers will not do business with an exchange partner they do not trust. As a first step toward a trust analysis procedure facilitating the design of a more trustworthy exchange site, this chapter defined a trust taxonomy enabling the classification of trust production methods in computer-mediated exchange. Applying the taxonomy to an exchange site enables Web designers to compare how that site is to be perceived against the classification of trust cues actually produced, identifying missed opportunities for trust production in terms of untapped sources and dimensions of trust. Although this work provides a promising first step toward a complete trust analysis procedure, much work still remains in assessing a user's perception of trust cues produced within the unique context of computer-mediated exchange.

REFERENCES

Bailey, B. P., Gurak, L. J., & Konstan, J. A. (2001). An examination of trust production in computer-mediated exchange. In *Proceedings of Human Factors and the Web*.

Brainov, S., & Sandholm, T. (1999). Contracting with uncertain level of trust. In *Proceedings of ACM Conference on Electronic Commerce* (pp. 15–21).

Deutsch, M. A. (1958). Trust and suspicion. *Journal of Conflict Resolution, 2*, 265–279.

Deutsch, M. A. (1973). *The resolution of conflict: Constructive and destructive processes*. New Heaven, CT: Yale University Press.

Doney, P. M., & Cannon, J. P. (1997). An examination of the nature of trust in buyer–seller relationships. *Journal of Marketing, 61*(2), 35–51.

Fogg, B. J. (1998). Persuasive computers: Perspectives and research directions. In *Proceedings of ACM Conference on Human Factors and Computing Systems* (pp. 225–232).

Fogg, B. J., & Tseng, H. (1999). The elements of computer credibility. In *Proceedings of the ACM Conference on Human Factors and Computing Systems* (pp. 80–87).

Giffin, K. (1967). The contribution of studies of source credibility to a theory of interpersonal trust in the communication process. *Psychological Bulletin, 68*(2), 104–120.

(1999). E-commerce trust study. [Online]. Available: http://www.cheskin.com/think/studies/ecomtrust. html

Johnson, D. S., & Grayson, K. (2000). Sources and dimensions of trust in service relationships. In D. Iacobucci & T. Swartz (Eds.), *Handbook of services marketing & management* (pp. 357–370). Thousand Oaks, CA: Sage.

Keen, P. (1997, April 21). Are you ready for the "trust" economy? *Computerworld, 31*(80).

Kim, J., & Moon, J. Y. (1997). Designing towards emotional usability in customer interfaces: Trustworthiness of cyber-banking system interfaces. *Interacting with Computers, 10*, 1–29.

Nass, C., Moon, Y., Fogg, B. J., Reeves, B., & Dryer, C. (1995). Can computer personalities be human personalities? *International Journal of Human-Computer Studies, 43*, 223–239.

Nass, C., Steuer, J., & Tauber, E. R. (1994). Computers are social actors. In *Proceedings of the ACM Conference on Human Factors and Computing Systems* (pp. 72–77).

Nass, C., Reeves, B., & Leshner, G. (1996). Technology and roles: A tale of two TVs. *Journal of Communication, 46*(2), 121–128.

Patzer, G. L. (1983, June). Source credibility as a function of communicator physical attractiveness. *Journal of Business Research, 11*(2), 229–241.

Peters, R. G., Covello, V. T., & McCallum, D. B. (1997, February). The determinants of trust and credibility in environmental risk communication: An empirical study. *Risk Analysis, 17*(1), 43–54.

Posner, B. Z., & Kouzes, J. M. (1998, October). Relating leadership and credibility. *Psychological Reports, 63*(2), 527–530.

Rempel, J. K., Holmes, J. G., & Zanna, M. P. (1985, July). Trust in close relationships. *Journal of Personality & Social Psychology, 49*(1), 95–112.

Rice, R. E., & Love, G. (1987, February). Electronic emotion: Socioemotional content in a computer-mediated communication network. *Communication Research, 14*(1), 85–108.

Rocco, E. (1998). Trust breaks down in electronic contexts but can be repaired by some initial face-to-face contact. In *Proceedings of ACM Conference on Human Factors and Computing Systems* (pp. 496–502).

Rotter, J. B. (1980, January). Interpersonal trust, trustworthiness, and gullibility. *American Psychologist, 35*(1), 1–7.

Schurr, P. H., & Ozanne, J. L. (1985). Influences on exchange processes: Buyers' preconceptions of a seller's trustworthiness and bargaining toughness. *Journal of Consumer Research, 11*(4), 939–953.

Sproull, L., & Kiesler, S. (1986). Reducing social context cues: Electronic mail in organizational communication. *Management Science, 32*(11), 1492–1512.

Tseng, S., & Fogg, B. J. (1999, May). Credibility and computing technology. *Communications of the ACM, 42*(5), 39–44.

Waern, Y., & Ramberg, R. (1996). People's perception of human and computer advice. *Computers in Human Behavior, 12*(1), 17–27.

Walther, J. B. (1992, September). Relational communication in computer-mediated interaction. *Human Communication Research, 19*(1), 50–88.

Zajonc, R. B. (1980, February). Feeling and thinking. *American Psychologist, 35*(2), 151–175.

Zucker, L. G. (1986). Production of trust: Institutional sources of economic structure, 1840–1920. *Research in Organizational Behavior, 8*, 53–111.

Author Index

Subject Index